Oberwolfach Seminars
Volume 42

Willy Dörfler
Armin Lechleiter
Michael Plum
Guido Schneider
Christian Wieners

Photonic Crystals: Mathematical Analysis and Numerical Approximation

 Birkhäuser

Willy Dörfler
Institut für Angewandte und Numerische
Mathematik 2
Karlsruher Institut für Technologie
76128 Karlsruhe
Germany
willy.doerfler@kit.edu

Michael Plum
Institut für Analysis
Karlsruher Institut für Technologie
76128 Karlsruhe
Germany
michael.plum@kit.edu

Christian Wieners
Institut für Angewandte und Numerische
Mathematik 3
Karlsruher Institut für Technologie
76128 Karlsruhe
Germany
christian.wieners@kit.edu

Armin Lechleiter
INRIA Saclay and
CMAP Ecole Polytechnique
Route de Saclay
91128 Palaiseau Cedex
France
alechle@cmap.polytechnique.fr

Guido Schneider
Fachbereich Mathematik
Universität Stuttgart
Pfaffenwaldring 57
70569 Stuttgart
Germany
guido.schneider@mathematik.uni-stuttgart.de

2010 Mathematics Subject Classification: 35B, 35Q, 35R, 78A

ISBN 978-3-0348-0112-6   e-ISBN 978-3-0348-0113-3
DOI 10.1007/978-3-0348-0113-3

Library of Congress Control Number: 2011926985

Cover design: deblik

Printed on acid-free paper

Springer Basel AG is part of Springer Science+Business Media

www.birkhauser-science.com

Contents

Preface

Nanotechnology is a key technology of the 21st century. It investigates very small structures of the size of a few nanometers up to several 100 nanometers. Thus, these structures are often smaller then the wave length of light. In this book, we concentrate on the mathematics of photonic crystals, which form an important class of physical structures investigated in nanotechnology.

The investigation of these structures by mathematical methods is highly important for the following reasons.

- Since the physical behaviour in the nanoscale is very difficult and expensive to measure in real experiments, numerical simulations play a fundamental role in understanding such processes. In many cases theses structures are fully three dimensional, and process on very different scales in space and time, so that constructing efficient and reliable simulation methods is a true mathematical challenge.

- Often ill-posed problems arise in a natural way, e. g., in the reconstruction problem of nanostructures from measurements. Here, methods of the theory of inverse problems must be developed and applied.

- Often it is not possible, e. g. due to very large differences in the underlying scales, to consider the full basic physical equations directly in a numerical simulation. Then, reduced and simplified models have to be constructed, and their analytical properties and approximation qualities have to be investigated.

- The numerical simulation can study only specific configurations. For a qualitative understanding of the behaviour of the underlying system, the mathematical analysis of the underlying equations is indispensable.

In the mathematical analysis and the numerical appromixation of the partial differential equations describing nanostructures, several mathematical difficulties arise, e. g., the appropriate treatment of nonlinearities, simultaneous occurrence of continuous and discrete spectrum, multiple scales in space and time, and the ill-posedness of these problems.

Photonic crystals are materials which are composed of two or more different dielectrics or metals, and which exhibit a spatially periodic structure, typically at the length scale of hundred nanometers. Photonic crystals can be fabricated using processes such as photolithography or vertical deposition methods. They also occur in nature, e. g. in the microscopic structure of certain bird feathers, butterfly wings, or beetle shells.

A characteristic feature of photonic crystals is that they strongly affect the propagation of light waves at certain optical frequencies. This is due to the fact that the optical density inside a photonic crystal varies periodically at the length scale of about 400 to 800 nanometers, i. e., precisely at the scale of the wavelengths of optical light waves. Light waves that penetrate a photonic crystal are therefore subject to periodic, multiple diffraction, which leads to coherent wave interference inside the crystal. Depending on the frequency of the incident light wave this interference can either be constructive or destructive. In the latter case the light wave is not able to propagate inside the photonic crystal. Typically, this phenomenon only occurs for bounded ranges of optical wave frequencies, if it does occur at all. Such a range of inhibited wave frequencies is called a photonic band gap. Light waves with frequencies inside a photonic band gap are totally reflected by the photonic crystal. It is this effect which causes, e. g., the iridescent colours of peacock feathers.

In the mathematical modeling of photonic crystals by Maxwell's equations with periodic permittivity, such photonic band gaps are described as gaps in the spectrum of a selfadjoint operator with periodic coefficients, while the frequency ranges where constructive interference takes place form the spectrum (which is arranged in bands) of this selfadjoint operator.

In this proceedings volume we collect a series of lectures which introduce into the mathematical background needed for the modeling and simulation of light, in particular in periodic media, and for its applications in optical devices. We start with an introduction to Maxwell's equations, which build the basis for the mathematical description of all electro-magnetic phenomena, and thus in particular of optical waves. Next, we focus on explicit methods for the numerical computation of photonic band gaps. Furthermore, a general introduction to the so-called Floquet–Bloch theory is given, which provides analytical tools to investigate the spectrum of periodic differential operators and provides the aforementioned band-gap structure of selfadjoint operators with periodic coefficients, such as they occur for Maxwell's equations in a photonic crystal. In the rest of this volume we consider two applications. In the first application the theory of direct and inverse scattering is introduced and applied to periodic media, and the second application investigates nonlinear optical effects in wave guides which can be described by the nonlinear Schrödinger equation.

We greatly appreciate the opportunity to give this course in the framework of the Oberwolfach seminars, and we would like to thank the Oberwolfach institute for their kind hospitality. Further, we want to thank for the kind assistance of Birkhäuser in the realisation of these lecture notes.

Chapter 1

Introduction

by Willy Dörfler

In this introduction we will present both the main physical concepts of electromagnetism (Section 1.1.1) and mathematical basic tools (Section 1.2) that are needed for the topics discussed here.

1.1 The Maxwell Equations

The foundations of electromagnetism were laid by James Clerk Maxwell [1] in the years 1861–1865. It not only provided insight into the nature of electromagnetism, but also gave the theory a clear mathematical structure. It consists of a system of partial differential equations and some constitutive laws that describe the interaction of the fields with matter. Here, we mainly consider waves of a fixed frequency (in the range of light) or pulses with concentrated frequencies. This leads to the simplification of the time-harmonic equations. On idealised periodic material arrangements one ends up with linear (and nonlinear) eigenvalue equations. Using results from spectral theory it is possible to understand the structure of the (linear) eigenvalue problems qualitatively.

1.1.1 The partial differential equations

We seek vector fields $D, E, B, H : \Omega \to \mathbb{R}^3$ such that *Maxwell equations*

$$\partial_t D - \nabla \times H = -J \qquad \text{(Ampère's circuital law)}, \tag{1.1}$$

$$\nabla \cdot D = \varrho \qquad \text{(Gauss's law)}, \tag{1.2}$$

$$\partial_t B + \nabla \times E = 0 \qquad \text{(Faraday's law of induction)}, \tag{1.3}$$

$$\nabla \cdot B = 0 \qquad \text{(Gauss's law for magnetism)}, \tag{1.4}$$

---

[1] 13 June 1831 – 5 November 1879

hold for given $\boldsymbol{J} : \Omega \to \mathbb{R}^3$ and $\varrho : \Omega \to \mathbb{R}$. We call $\boldsymbol{E}$ *electric field*, $\boldsymbol{D}$ *electric displacement field* (or *electric flux density*), $\boldsymbol{H}$ *magnetic field intensity*, $\boldsymbol{B}$ *magnetic induction*, and $\boldsymbol{J}$ and ϱ *electric current intensity* and *electric charge density*, respectively. Furthermore, there are *constitutive relations* $\boldsymbol{D}(\boldsymbol{E}, \boldsymbol{H})$ and $\boldsymbol{B}(\boldsymbol{E}, \boldsymbol{H})$ that describe the interaction of the fields with matter. Examples are given in Section 1.1.3. This yields 14 scalar equations for 12 unknown functions. From the above equations we can derive the compatibility condition

$$\partial_t \varrho + \boldsymbol{\nabla} \cdot \boldsymbol{J} = 0,$$

which expresses *conservation of charge*. For a physical interpretation of equations (1.1)–(1.4) see [17, Ch. 1].

1.1.2   The integral formulation

Let M be a smooth oriented two-dimensional manifold in $\mathbb{R}^3$ and $V \subset \mathbb{R}^3$ a bounded domain with smooth boundary. Using the well-known integral relations for sufficiently smooth vector fields $\boldsymbol{A} : V \to \mathbb{R}^3$, resp. $\boldsymbol{A} : M \to \mathbb{R}^3$,

$$\int_V \boldsymbol{\nabla} \cdot \boldsymbol{A} = \int_{\partial V} \boldsymbol{A} \cdot \boldsymbol{n}_V \quad (\textit{Gauss' theorem}),$$

$$\int_M (\boldsymbol{\nabla} \times \boldsymbol{A}) \cdot \boldsymbol{n}_M = \oint_{\partial M} \boldsymbol{A} \cdot \boldsymbol{t}_M \quad (\textit{Stokes' theorem}),$$

we can derive the following identities from (1.1)–(1.4)

$$\oint_{\partial M} \boldsymbol{H} \cdot \boldsymbol{t}_M = \int_M (\partial_t \boldsymbol{D} + \boldsymbol{J}) \cdot \boldsymbol{n}_M, \qquad \oint_{\partial V} \boldsymbol{D} \cdot \boldsymbol{n}_V = \int_V \varrho, \qquad (1.5)$$

$$\oint_{\partial M} \boldsymbol{E} \cdot \boldsymbol{t}_M = -\int_M \partial_t \boldsymbol{B} \cdot \boldsymbol{n}_M, \qquad \oint_{\partial V} \boldsymbol{B} \cdot \boldsymbol{n}_V = 0. \qquad (1.6)$$

Here, $\boldsymbol{n}_M$ is the normal vector field on M, $\boldsymbol{t}_M$ is the tangential vector field on ∂M (oriented counter-clockwise if one looks into the direction of $\boldsymbol{n}_M$), and $\boldsymbol{n}_V$ the exterior normal to V.

1.1.3   Constitutive relations

In vacuum there hold the relations

$$\boldsymbol{D}(t, \boldsymbol{x}) = \varepsilon_0 \boldsymbol{E}(t, \boldsymbol{x}), \qquad \boldsymbol{B}(t, \boldsymbol{x}) = \mu_0 \boldsymbol{H}(t, \boldsymbol{x})$$

with the *permittivity of free space* ε_0 and the *permeability of free space* μ_0. In matter, however, the fields have to be interpreted in a macroscopic way as a mean field. An electric field $\boldsymbol{E}$ induces local dipoles in nonconducting media by dislocation of charges (as sketched in Figure 1.1). This gives rise to a *polarisation field* $\boldsymbol{P}$, that superposes the electric field and results in a total field

$$\boldsymbol{D} = \varepsilon_0 \boldsymbol{E} + \boldsymbol{P}.$$

The connection between $\boldsymbol{E}$ und $\boldsymbol{P}$ can be described by a material model. For a *linear medium* one makes the ansatz

$$\boldsymbol{P}(t, \boldsymbol{x}) = \varepsilon_0 (\boldsymbol{G} * \boldsymbol{E})(t, \boldsymbol{x}) := \varepsilon_0 \int_{-\infty}^{\infty} \boldsymbol{G}(t - \tau, \boldsymbol{x}) \boldsymbol{E}(\tau, \boldsymbol{x}) \, d\tau. \tag{1.7}$$

Figure 1.1: Dislocation of local charges by an exterior electric field $\boldsymbol{E}$ and the polarisation field $\boldsymbol{P}$ (without (left) and with (right) exterior electrical field).

The *transfer function* $\boldsymbol{G}(\cdot, \boldsymbol{x}) : \mathbb{R} \to \mathbb{R}^{3,3}$ is assumed to be *causal*, which means that $\boldsymbol{P}(t, \, . \,)$ can only depend on $\boldsymbol{E}(\tau, \, . \,)$ on past times $\tau \leq t$. This is expressed by the requirement

$$\boldsymbol{G}(s, \boldsymbol{x}) = \boldsymbol{0} \qquad \text{for all } s < 0 \text{ and all } \boldsymbol{x} \in \mathbb{R}^3.$$

Special cases are *homogeneous* media, where $\boldsymbol{G}$ does not depend on $\boldsymbol{x}$ and *isotropic* media, where $\boldsymbol{G}$ is real valued. If we perform a *Fourier transformation* in time, e. g.,

$$\widehat{\boldsymbol{D}}(\omega, \boldsymbol{x}) = \int_{\mathbb{R}^3} \boldsymbol{D}(t, \boldsymbol{x}) e^{-i\omega t} \, dt$$

(and likewise for $\boldsymbol{E}, \boldsymbol{P}$), the convolution will turn into

$$\widehat{\boldsymbol{P}}(\omega, \boldsymbol{x}) = \varepsilon_0 \chi_{\mathrm{E}}(\omega, \boldsymbol{x}) \widehat{\boldsymbol{E}}(\omega, \boldsymbol{x})$$

with the *electric susceptibility*

$$\chi_{\mathrm{E}}(\omega, \boldsymbol{x}) = \frac{1}{\epsilon_0} \widehat{\boldsymbol{G}}(\omega, \boldsymbol{x}).$$

Thus we obtain the relation

$$\widehat{\boldsymbol{D}}(\omega, \boldsymbol{x}) = \varepsilon_0 \big(1 + \chi_{\mathrm{E}}(\omega, \boldsymbol{x}) \big) \widehat{\boldsymbol{E}}(\omega, \boldsymbol{x})$$
$$\equiv \varepsilon_0 \varepsilon_{\mathrm{r}}(\omega, \boldsymbol{x}) \widehat{\boldsymbol{E}}(\omega, \boldsymbol{x}). \tag{1.8}$$

Note that a relation $\boldsymbol{D} = \varepsilon_0\varepsilon_r\boldsymbol{E}$ will only follow if the material parameter ε_r is frequency independent, or, if the relation is considered at a single frequency ω_0. To see this, multiply (1.8) by the Dirac measure δ_{ω_0} and apply the inverse Fourier transform. This independency is not given in reality, however, it often holds approximately on large frequency intervals. Linear relations $\boldsymbol{B} = \mu_0(\boldsymbol{H}+\boldsymbol{M})$, $\boldsymbol{M}$ the *magnetisation*, and $\boldsymbol{B} = \mu_0\mu_r\boldsymbol{H}$ follows from analogous reasoning.

Usually, ε_r is taken to be real. This can only be an approximation, since ε_r *must* have an imaginary part, except in the vacuum case, by the *Kramers–Kronig relation*, that relates $\Im(\varepsilon_r)$ to $\Re(\varepsilon_r) - 1$ [17, Ch. 7.10].

In the following, we will use the linear constitutive relations

$$\boldsymbol{D} = \varepsilon\boldsymbol{E} = \varepsilon_0\varepsilon_r\boldsymbol{E}, \tag{1.9}$$

$$\boldsymbol{B} = \mu\boldsymbol{H} = \mu_0\mu_r\boldsymbol{H} \tag{1.10}$$

(here with possibly space-dependent but time-independent scalar functions ε, μ). ε, μ (ε_r, μ_r) are called *(relative) permittivity* and *(relative) permeability*, respectively.

The interaction of electromagnetic fields with media is a rich and still developing topic. We mention only a few models.

The homogeneous, isotropic one–resonance model

Here it is assumed that the mentioned local dipoles are oscillating systems at resonance frequency ω_0, that oscillate in the direction of the electric field. In this model the equation for $\boldsymbol{P}$ is that of a *damped harmonic oscillator* and can be solved explicitly. As a result one obtains the *Drude–Lorentz model*

$$\chi_E(\omega) = \frac{1}{\epsilon_0}\frac{\omega_p^2}{\omega_0^2 - \omega^2 - i\gamma\omega},$$

where ω_p^2 and γ are material constants named *plasma frequency* and *damping factor*, respectively [17, Ch. 7.5]. The dependence of ϵ on ω is especially important for metallic materials. Here, one often uses the expression with $\omega_0 := 0$ (*Drude model*).

A quantum mechanical model

In a medium or an a scale where quantum effects become apparent the polarisation will depend on a *density matrix* $\varrho \in \mathbb{C}^{N,N}$ with a number N of energy levels. ϱ is hermitian and nonnegative with positive diagonal entries. Then one finds a dependence $\boldsymbol{P}(\boldsymbol{E}, \varrho)$ and ϱ satisfies an equation of the form $\partial_t\varrho = \boldsymbol{L}(\boldsymbol{E}, \varrho)$, where $\boldsymbol{L}$ is affine linear in both arguments. The final set of (nonlinear) equations is called *Maxwell–Bloch equations*. In case of $N = 2$ (*two level model*) one arrives for example at equations

$$\partial_t^2 \boldsymbol{P} + \alpha_1 \partial_t \boldsymbol{P} + \omega^2 \boldsymbol{P} = \alpha_2 \varrho \boldsymbol{E},$$
$$\partial_t \varrho + \alpha_3(\varrho - \varrho_0) = -\alpha_4 \partial_t \boldsymbol{P} \cdot \boldsymbol{E}.$$

$\alpha_1, \ldots, \alpha_4$ are positive constants and ϱ is the difference in number between excited states and ground states per unit volume [22]. For mathematical investigations see [8] [9].

Effective medium approximations

These are methods to describe the effective permittivity or effective permeability of composed materials when the material structures are very small, like the *Bruggeman model* [4] or the *Maxwell–Garnett model* [15]. Especially newly designed materials at nano-scale with electric and magnetic effects can result in effective media with unusual optical properties (*meta materials*) [23].

Nonlinear polarisation

The previous approach (1.7) can be generalised to higher polynomial dependence on $\boldsymbol{E}$ [1]. Of special importance is the cubic (*Kerr–*) nonlinearity

$$\boldsymbol{P} = \varepsilon_0(\varepsilon_r + \alpha|\boldsymbol{E}|^2)\boldsymbol{E}$$

(with some parameter $\alpha \in \mathbb{R}$). Equations with cubic nonlinearities will be studied in Chapter 5.

1.1.4 The wave equation

We consider the case of an electromagnetic field in the absence of charges and currents ($\varrho = 0$ and $\boldsymbol{J} = \boldsymbol{0}$) and in the presence of a material with constitutive relations $\boldsymbol{D} = \varepsilon_0 \boldsymbol{E} + \boldsymbol{P}$ and (1.10). From the Maxwell equations (1.1) and (1.3) we get

$$\partial_t^2(\varepsilon_0 \boldsymbol{E} + \boldsymbol{P}) = \partial_t^2 \boldsymbol{D} = \partial_t(\boldsymbol{\nabla} \times \boldsymbol{H}) = -\boldsymbol{\nabla} \times \left(\frac{1}{\mu} \boldsymbol{\nabla} \times \boldsymbol{E}\right)$$

or
$$\varepsilon_0 \partial_t^2 \left(\boldsymbol{E} + \frac{1}{\varepsilon_0} \boldsymbol{P}\right) + \boldsymbol{\nabla} \times \left(\frac{1}{\mu} \boldsymbol{\nabla} \times \boldsymbol{E}\right) = 0.$$

Using the linear relations (1.8) and (1.9), we find $\boldsymbol{E} + 1/\varepsilon_0 \, \boldsymbol{P} = \varepsilon \boldsymbol{E}$ and therefore

$$\varepsilon \partial_t^2 \boldsymbol{E} + \boldsymbol{\nabla} \times \left(\frac{1}{\mu} \boldsymbol{\nabla} \times \boldsymbol{E}\right) = 0.$$

In vacuum, we have $\varepsilon = \varepsilon_0$, $\mu = \mu_0$. With $0 = \boldsymbol{\nabla} \cdot (\varepsilon_0 \boldsymbol{E}) = \varepsilon_0 \boldsymbol{\nabla} \cdot \boldsymbol{E}$ we observe that

$$\boldsymbol{\nabla} \times (\boldsymbol{\nabla} \times \boldsymbol{E}) = \boldsymbol{\nabla}(\boldsymbol{\nabla} \cdot \boldsymbol{E}) - \Delta \boldsymbol{E} = -\Delta \boldsymbol{E}$$

and thus obtain the *wave equation*

$$\frac{1}{c^2}\partial_t^2 \boldsymbol{E} - \Delta \boldsymbol{E} = \boldsymbol{0}, \tag{1.11}$$

where $c^2 = 1/(\varepsilon_0\mu_0)$ is the *speed of light*. An analogous equation holds for $\boldsymbol{H}$. Note, that the wave equation has solutions of the form $\boldsymbol{E}(t, \boldsymbol{x}) = \boldsymbol{A}(\boldsymbol{x} \pm ct\boldsymbol{n})$ for some suitable function $\boldsymbol{A} : \mathbb{R}^3 \to \mathbb{R}^3$ and a unit vector $\boldsymbol{n}$. This models the propagation of light in vacuum.

1.1.5 Time–harmonic Maxwell equations

In case of a monochromatic wave we can assume that all fields are of the form $\boldsymbol{E}(\boldsymbol{x})e^{i\omega t}$, etc. With $\varrho = 0$ and $\boldsymbol{J} = \boldsymbol{0}$ and the linear relations $\boldsymbol{D} = \varepsilon\boldsymbol{E}$ and $\boldsymbol{B} = \mu\boldsymbol{H}$, the Maxwell equations (1.1)–(1.4) turn into the *time-harmonic Maxwell equations*

$$i\omega\varepsilon\boldsymbol{E} - \nabla \times \boldsymbol{H} = \boldsymbol{0}, \tag{1.12}$$

$$\nabla\cdot(\varepsilon\boldsymbol{E}) = 0, \tag{1.13}$$

$$i\omega\mu\boldsymbol{H} + \nabla \times \boldsymbol{E} = \boldsymbol{0}, \tag{1.14}$$

$$\nabla\cdot(\mu\boldsymbol{H}) = 0. \tag{1.15}$$

Elimination of $\boldsymbol{E}$ yields

$$\nabla \times \left(\frac{1}{\varepsilon}\nabla \times \boldsymbol{H}\right) = \nabla \times (i\omega\boldsymbol{E}) = i\omega(-i\omega)\mu\boldsymbol{H} = \omega^2\mu\boldsymbol{H}, \tag{1.16}$$

while elimination of $\boldsymbol{H}$ gives analogously

$$\nabla \times \left(\frac{1}{\mu}\nabla \times \boldsymbol{E}\right) = \omega^2\varepsilon\boldsymbol{E}. \tag{1.17}$$

We thus finally arrive at *eigenvalue problems* for $\boldsymbol{H}$ and $\boldsymbol{E}$. It means that, except in cases where the frequency ω (actually, ω^2) is an eigenvalue of this equation, there exists no nontrivial electric field and thus no wave is transmitted. Note that a solution of one of this problems will determine a solution of the other, using the first order equations (1.12) and (1.14) above. Note further that (1.16) and (1.17) have to be accompanied by (1.13) and (1.15), respectively, but formally, the ladder follow as a consequence from the previous by taking the divergence and using $\nabla\cdot\nabla\times = 0$.

In the following we study this problem in special geometric configurations for illustration.

1d–like structures

Assume that we have a layered structure in x_1–direction, with material properties $x_1 \mapsto \varepsilon(x_1)$ and $x_1 \mapsto \mu(x_1)$ (see Figure 1.2 (left)). We especially seek an electric

Figure 1.2: Examples of a one-dimensional (left) and two-dimensional (right) structure.

field of the form $x_1 \mapsto \boldsymbol{E}(x_1)$. As an example, we think of an idealisation of a device that consists of plane layers of material with alternating constant permittivity and that we look for nonvanishing standing waves $\boldsymbol{E}(x_1)e^{i\omega t}$ for $\omega \neq 0$. Note that in this case

$$\boldsymbol{\nabla} \times \boldsymbol{E} = \begin{bmatrix} 0 \\ -\partial_1 E_3 \\ \partial_1 E_2 \end{bmatrix}$$

and therefore

$$\boldsymbol{\nabla} \times \left(\frac{1}{\mu} \begin{bmatrix} 0 \\ -\partial_1 E_3 \\ \partial_1 E_2 \end{bmatrix} \right) = - \begin{bmatrix} 0 \\ \partial_1 \left(\frac{1}{\mu} \partial_1 E_2 \right) \\ \partial_1 \left(\frac{1}{\mu} \partial_1 E_3 \right) \end{bmatrix}.$$

As a consequence we obtain from (1.17) the decoupled system

$$E_1 = 0,$$

$$-\partial_1 \left(\frac{1}{\mu} \partial_1 E_j \right) = \omega^2 \varepsilon E_j \quad \text{for } j = 2, 3.$$

The equation for $\boldsymbol{H}$ is analogous with ε and μ exchanged. As a result, we obtain standard elliptic eigenvalue problems for E_2 and E_3.

2d–like structures (waveguide)

Consider now a material that consists of infinite columns in x_3–direction modeled by a permittivity $(x_1, x_2) \mapsto \varepsilon(x_1, x_2)$ (see Figure 1.2 (right)), and we look for time-harmonic solutions that are also periodic in x_3–direction along the structure. Thus we seek fields $(t, x_1, x_2, x_3) \mapsto \boldsymbol{E}(x_1, x_2)e^{ik_3 x_3}e^{i\omega t}$ and $(t, x_1, x_2, x_3) \mapsto \boldsymbol{H}(x_1, x_2)e^{ik_3 x_3}e^{i\omega t}$ for $\omega \neq 0$. This corresponds to a wave with frequency ω that travels in x_3–direction with wavenumber k_3. This structure is called a *waveguide* if nontrivial fields of this kind exist. For such a vector field $\boldsymbol{E}$ we obtain

$$\boldsymbol{\nabla} \times (\boldsymbol{E}e^{ik_3 x_3}) = \begin{bmatrix} \boldsymbol{\nabla}^{\perp}_{1,2} E_3 - ik_3[E_1; E_2]^{\perp} \\ -\boldsymbol{\nabla}^{\perp}_{1,2} \cdot [E_1; E_2] \end{bmatrix} e^{ik_3 x_3},$$

where $\boldsymbol{\nabla}_{1,2} := [\partial_1; \partial_2]$, $\boldsymbol{\nabla}_{1,2}^{\perp} := [\partial_2; -\partial_1]$, and $[E_1; E_2]^{\perp} = [E_2; -E_1]$. If we insert this into (1.17), we get a system that can be separated as

$$-\boldsymbol{\nabla}_{1,2}^{\perp}\left(\frac{1}{\mu}\boldsymbol{\nabla}_{1,2}^{\perp} \cdot \begin{bmatrix} E_1 \\ E_2 \end{bmatrix}\right) + i\frac{1}{\mu}k_3\boldsymbol{\nabla}_{1,2}E_3 + \frac{1}{\mu}k_3^2\begin{bmatrix} E_1 \\ E_2 \end{bmatrix} = \omega^2\varepsilon\begin{bmatrix} E_1 \\ E_2 \end{bmatrix}, \quad (1.18)$$

$$-\boldsymbol{\nabla}_{1,2} \cdot \left(\frac{1}{\mu}\boldsymbol{\nabla}_{1,2}E_3\right) + i\frac{1}{\mu}k_3\boldsymbol{\nabla}_{1,2} \cdot \begin{bmatrix} E_1 \\ E_2 \end{bmatrix} = \omega^2\varepsilon \ E_3. \quad (1.19)$$

For given k_3 this is an eigenvalue problem for ω, while for given ω this is a quadratic eigenvalue problem for k_3 (for such problems see [25]). In case of $k_3 = 0$ (standing wave) this system decouples as

$$-\boldsymbol{\nabla}_{1,2}^{\perp}\left(\frac{1}{\mu}\boldsymbol{\nabla}_{1,2}^{\perp} \cdot \begin{bmatrix} E_1 \\ E_2 \end{bmatrix}\right) = \omega^2\varepsilon\begin{bmatrix} E_1 \\ E_2 \end{bmatrix},$$

$$-\boldsymbol{\nabla}_{1,2} \cdot \left(\frac{1}{\mu}\boldsymbol{\nabla}_{1,2}E_3\right) = \omega^2\varepsilon E_3.$$

The second equation is an elliptic eigenvalue problem for (E_3, ω). Correspondingly, one can also obtain (H_3, ω) by the same type of equation (but with μ and ε interchanged). Having determined in this way E_3 and H_3, we derive equations for the remaining components from Maxwell's first order equations

$$\begin{bmatrix} E_1 \\ E_2 \end{bmatrix} = \frac{1}{i\omega\varepsilon}\boldsymbol{\nabla}_{1,2}^{\perp}H_3, \qquad \begin{bmatrix} H_1 \\ H_2 \end{bmatrix} = -\frac{1}{i\omega\mu}\boldsymbol{\nabla}_{1,2}^{\perp}E_3.$$

Waves that are described by the components E_1, E_2, H_3 are called *transverse-electric* (TE) and those described by the components H_1, H_2, E_3 are called *transverse-magnetic* (TM). If μ and ε are constant, then (1.18)–(1.19) simplifies to the eigenvalue problem

$$-\Delta_{1,2}\boldsymbol{E} = \left(\omega^2\varepsilon\mu - k_3^2\right)\boldsymbol{E}.$$

However, this result could easier be obtained from (1.11).

We have seen that problems with 1D or 2D structures can be solved using an eigenvalue problem for the Laplace-operator. The *off-plane* problem ($k_3 \neq 0$) however leads to the more involved equation (1.18). Also, full 3D problems will in general be of a more complex type.

The construction of waveguides is an important topic in optical communication technology [1], see also Chapter 5.7.3.

1.1.6   Boundary and interface conditions

Electromagnetic fields generally exist in the whole space, they are generated by charges and currents (ϱ and $\boldsymbol{J}$ in (1.1) and (1.2)) and influenced by interaction with matter as indicated in Section 1.1.3. In our macroscopic modeling, permittivity and permeability are assumed to be smooth functions that are discontinuous along smooth material interfaces. It is therefore important to consider how this influences the electromagnetic field.

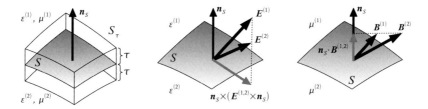

Figure 1.3: Interface conditions: Neighbourhood of the interface S (left), continuity of the tangential electric field component (middle), continuity of the normal electric field component (right).

Jump conditions

Let S be a surface, or part of a surface, where material parameters jump. If we consider the cylindrical domain $S_\tau = S \times (-\tau, \tau)$ for $\tau > 0$ (see Figure 1.3) and if a *surface charge density* σ_S is located on S, then we obtain in the limit $\tau \searrow 0$ from (1.5)

$$\int_S (\boldsymbol{D}^{(1)} - \boldsymbol{D}^{(2)}) \cdot \boldsymbol{n}_S = \sigma_S \operatorname{vol}(S).$$

Since S is arbitrary, we conclude the pointwise equation

$$(\boldsymbol{D}^{(1)} - \boldsymbol{D}^{(2)}) \cdot \boldsymbol{n}_S = \sigma_S \quad \text{on } S.$$

Applying this idea to the other Maxwell equations in Section 1.1.2 yields that

$$\boldsymbol{n}_S \cdot (\boldsymbol{B}^{(1)} - \boldsymbol{B}^{(2)}) = 0,$$
$$\boldsymbol{n}_S \times (\boldsymbol{E}^{(1)} - \boldsymbol{E}^{(2)}) = 0,$$
$$\boldsymbol{n}_S \times (\boldsymbol{H}^{(1)} - \boldsymbol{H}^{(2)}) = \boldsymbol{J}_S,$$

holds on S, where $\boldsymbol{J}_S$ is a *surface current density*.

Perfect conductors

In a conductor an electric field gives rise to an electric current by *Ohm's law* $\boldsymbol{J} = \sigma \boldsymbol{E}$ for some $\sigma > 0$. σ is called *electrical conductivity*. Good conductors, like metals, have large values of σ. A *perfect conductor* is a material that is characterised by the formal limit $\sigma \to \infty$. In the time-harmonic case of Section 1.1.5 we get in this limit

$$\boldsymbol{E} = \frac{1}{\sigma}\left(\mathrm{i}\omega\varepsilon\boldsymbol{E} + \nabla \times \boldsymbol{H}\right) \to \boldsymbol{0}.$$

Therefore the electric field, and thus the magnetic field, vanish in a perfect conductor. If one side of S is occupied by a perfect conductor, we thus get the condition

$$\boldsymbol{n}_S \times \boldsymbol{E} = \boldsymbol{0} \text{ and } \boldsymbol{n}_S \cdot \boldsymbol{B} = 0 \quad \text{on } S \tag{1.20}$$

for the fields $\boldsymbol{E}$ and $\boldsymbol{B}$ on the other side from the jump conditions. Note that if $\boldsymbol{E}$ is a potential, $\boldsymbol{E} = -\boldsymbol{\nabla}\Psi$, then $\boldsymbol{n}_S \times \boldsymbol{E} = \boldsymbol{0}$ on S implies that the tangential derivative of Ψ vanishes on S, meaning that S is a level set for Ψ.

Unbounded domains

On an unbounded domain one needs conditions for $|\boldsymbol{x}| \to \infty$. We split the electric field $\boldsymbol{E}$ into the incident part, e. g. a plane wave $\boldsymbol{E}^i(\boldsymbol{x}) = e^{i\boldsymbol{k}\cdot\boldsymbol{x}}$, and a part $\boldsymbol{E}^s$ that is scattered by some object, $\boldsymbol{E} = \boldsymbol{E}^i + \boldsymbol{E}^s$. For the scattered wave we impose the *Silver–Müller radiation condition*

$$\lim_{|\boldsymbol{x}|\to\infty} |\boldsymbol{x}|\left((\boldsymbol{\nabla}\times\boldsymbol{E}^s)(\boldsymbol{x})\times\frac{\boldsymbol{x}}{|\boldsymbol{x}|} - \mathrm{i}\omega\boldsymbol{E}^s(\boldsymbol{x})\right) = \boldsymbol{0}.$$

This means that there is no wave emanating from ∞. This topic will be studied in more detail in Chapters 4.1.1 and 4.1.2.

1.1.7   Photonic bandstructure

Our purpose is to study the propagation of electromagnetic waves in crystals. A crystal is a periodic configuration of atoms or molecules, here idealised to exist in the whole space of $\mathbb{R}^d$. It is modeled by a periodic permittivity function ε_r and permeability μ_r, more precisely, assume that Γ is a lattice spanned by the linearly independent directions $\boldsymbol{a}^1, \boldsymbol{a}^2, \ldots, \boldsymbol{a}^d \in \mathbb{R}^d$,

$$\Gamma := \left\{ \sum_{j=1}^{d} n_j \boldsymbol{a}^j \ : \ n_j \in \mathbb{Z}, \ j = 1, \ldots, d \right\},$$

and that ε_r and μ_r are periodic functions on Γ, that is,

$$\varepsilon_r(\boldsymbol{x} + \boldsymbol{R}) = \varepsilon_r(\boldsymbol{x}), \ \mu_r(\boldsymbol{x} + \boldsymbol{R}) = \mu_r(\boldsymbol{x}) \quad \text{for all } \boldsymbol{x} \in \mathbb{R}^d, \ \boldsymbol{R} \in \Gamma.$$

In the following we will assume the special case that $\boldsymbol{a}^i = \boldsymbol{e}^i$, the i-th euclidian unit vector.

The problem we want to consider is the eigenvalue problem (1.17) (or (1.16)), which can be written in the form

$$L(\boldsymbol{\nabla})\boldsymbol{E} = \lambda\varepsilon_r\boldsymbol{E} \quad \text{in } \mathbb{R}^d,$$

where L is linear differential operator with periodic coefficients. Physically, such eigenfunctions cannot exist since they have unbounded energy. Nevertheless, we see by *Bloch–Floquet theory* in Chapter 3 that the full spectral information for the corresponding Maxwell operator in $\mathbb{R}^3$ is obtained from *quasi-periodic* eigenfunctions, that is, eigenfunctions that satisfy

$$\boldsymbol{E_k}(\boldsymbol{x} + \boldsymbol{R}) = \exp\big(\mathrm{i}\boldsymbol{k} \cdot \boldsymbol{R}\big)\boldsymbol{E_k}(\boldsymbol{x}) \tag{1.21}$$

for all $\boldsymbol{x} \in \mathbb{R}^d$, $\boldsymbol{R} \in \Gamma$ and $\boldsymbol{k} \in \mathbb{R}^2$. If we define

$$\boldsymbol{u_k}(\boldsymbol{x}) := \exp\big(-\mathrm{i}\boldsymbol{k} \cdot \boldsymbol{x}\big)\boldsymbol{E_k}(\boldsymbol{x}),$$

we get that $\boldsymbol{u_k}$ is periodic on Γ,

$$\boldsymbol{u_k}(\boldsymbol{x} + \boldsymbol{R}) = \boldsymbol{u_k}(\boldsymbol{x}) \quad \text{for all } \boldsymbol{x} \in \mathbb{R}^d, \, \boldsymbol{R} \in \Gamma.$$

Consequently, $\boldsymbol{u_k}$ is completely determined by its values on the *fundamental cell* W of the lattice Γ,

$$W := \Big\{ \sum_{j=1}^{d} s_j \boldsymbol{a}^j \, : \, s_j \in \big[-\tfrac{1}{2}, \tfrac{1}{2}\big], \, j = 1, \ldots, d \Big\} \qquad (\textit{Wigner–Seitz cell}),$$

(see Figure 1.4 (left)) and thus the equation for $\boldsymbol{E_k}$ turns into the eigenvalue equation

$$L(\boldsymbol{\nabla} + \mathrm{i}\boldsymbol{k})\boldsymbol{u_k} = \lambda \boldsymbol{u_k} \quad \text{in } W \tag{1.22}$$

for $\boldsymbol{u_k}$. Moreover, if we define the *dual grid*

$$\Gamma^* := \Big\{ 2\pi \sum_{j=1}^{d} n_j \boldsymbol{b}^j \, : \, n_j \in \mathbb{Z}, \, j = 1, \ldots, d \Big\},$$

where $\boldsymbol{b}^1, \ldots, \boldsymbol{b}^d$ are unit vectors such that $\boldsymbol{b}^l \cdot \boldsymbol{a}^m = \delta_{lm}$ for $l, m \in \{1, \ldots, d\}$ (in our setting $\boldsymbol{b}^j = \boldsymbol{e}^j$), we observe that $\boldsymbol{E_{k+K}}$ also satisfies (1.21) like $\boldsymbol{E_k}$, since $\exp(-\mathrm{i}(\boldsymbol{k} + \boldsymbol{K}) \cdot \boldsymbol{x}) = \exp(-\mathrm{i}\boldsymbol{k} \cdot \boldsymbol{x})$ due to $\boldsymbol{K} \cdot \boldsymbol{R} \in 2\pi\mathbb{Z}$ for $\boldsymbol{R} \in \Gamma$, $\boldsymbol{K} \in \Gamma^*$. Thus it suffices to consider $\boldsymbol{k}$ in a fundamental cell B of Γ^*, e. g.,

$$B := \Big\{ \sum_{j=1}^{d} s_j \boldsymbol{b}^j \, : \, s_j \in [-\pi, \pi], \, j = 1, \ldots, d \Big\} \qquad (\textit{Brillouin zone})$$

(see Figure 1.4 (right)).

The essential theoretical result is that, for real and strictly positive ε_r, μ_r and fixed $\boldsymbol{k} \in B$, the reduced problem (1.22) is an elliptic eigenvalue problem on a bounded domain (corresponding to a self-adjoint operator with a compact inverse), so that the spectrum is discrete with real nonnegative eigenvalues $\lambda_{\boldsymbol{k},n}$

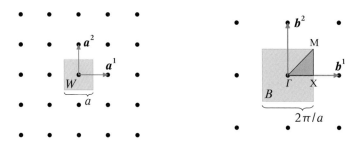

Figure 1.4: Orthogonal lattice (of size a) with Wigner–Seitz cell W (left), dual lattice with Brillouin zone B, the fundamental region and the high symmetry points (right).

and eigenfunctions $u_{k,n}$ for $n \in \mathbb{N}$ (see Chapter 2.1 (Theorem 2.1.7) and Chapter 3 for a more general theory). It is convenient to assume that they are ordered as

$$0 \leq \lambda_{k,1} \leq \lambda_{k,2} \leq \dots$$

for each k. The collection of curves $k \to \lambda_{k,n}$ is called the *photonic bandstructure* (see Figure 1.5). The curves are continuous by a result of [20, Ch. IV.2.6] (see also [7]). The total spectrum σ of the operator L is then the projection of all these curves onto the λ-axis, which can be expressed as a collection of intervals,

$$\sigma = \bigcup_{n \in \mathbb{N}} \Big[\inf_{k \in B} \lambda_{k,n}, \sup_{k \in B} \lambda_{k,n} \Big].$$

If for some $n \in \mathbb{N}$ the interval

$$\Big(\sup_{k \in B} \lambda_{k,n}, \inf_{k \in B} \lambda_{k,n+1} \Big)$$

is non-empty, this is called a *band gap*. In general, such periodic structures have no band gaps, and the existence of band gaps strongly depends on the material distribution. Analytically, only in very specific cases the existence of band gaps can be verified analytically [12] [13]. A rigourous computer-assisted proof for a band gap has been given in [16]. Thus, without physical experiments the prediction of band gaps require numerical methods.

In the more general case that ε_r depends also on ω, one ends up with a nonlinear eigenvalue problem $L(\nabla)E = \lambda \varepsilon_r(. , \omega)E$. Bandstructure computations in this case have been obtained in [10].

Molding the way of light

Bandgaps are interesting for the following reasons. They provide a range of frequencies, for which no propagated wave exists in the crystal. If one then modifies

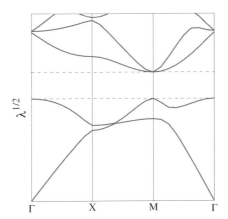

Figure 1.5: Photonic bandstructure of the 5 smallest eigenvalues along the path Γ, X, M, Γ (the high symmetry points, see Figure 1.4) around the Brillouin zone B. There is a bandgap between the second and third band.

the crystal along a path in a way that allows light propagation for band gap frequencies, the light is confined to this path. In this way one can construct circuits for light similar to those known from semiconductors [18] [14].

A challenging topic is to find materials with large bandgaps and to fabricate them. In a computational model one may start with some material and to enlarge bandgaps by varying the distribution of the permittivity. Such optimisation methods have been developed in [7] and [19]. This is especially challenging for 3D since in each optimisation step a full 3D Maxwell band structure has to be computed. Recent progress in this area has been made in [5] (see Chapter 2.2.12) and [24].

Acknowledgement. The author thanks Markus Richter for discussions an for providing figures from his thesis [24].

1.2 Some tools from analysis

1.2.1 Classical– and Sobolev–Function spaces

Here we will use the classical and the *Lebesgue* and *Sobolev* function spaces

$$C^0(\overline{\Omega}),\ C^{0,\alpha}(\overline{\Omega}),\ C^k(\overline{\Omega}),\ C^{k,\alpha}(\overline{\Omega}),$$
$$L^p(\Omega),\ W^{k,p}(\Omega),\ H^k(\Omega),\ W^{k,p}_0(\Omega),\ H^k_0(\Omega)$$

for $k \in \mathbb{N}$, $\alpha \in [0,1]$, $p \in [1,\infty]$, and their corresponding norms. Note, $C^0 \equiv C^{0,0}$, $W^{0,p} \equiv L^p$, and $H^k \equiv W^{k,2}$. In the following we let $\Omega \subset \mathbb{R}^n$ be a bounded domain with Lipschitz-continuous boundary. The *dual spaces* to $W^{k,p}_0(\Omega)$ are denoted by $W^{-k,p'}(\Omega)$ $(1/p + 1/p' = 1)$, analogously $H^{-k}(\Omega)$. Here, the norm is the operator norm of the functionals. The definitions extend to classes of functions $v : \Omega \subset \mathbb{R}^n \to \mathbb{R}^d$ (or $\mathbb{C}^d$) componentwise and by taking, for example, the euclidian vector norm. The notation will then be $W^{k,p}(\Omega)^d$. Existence of *continuous embeddings* between these spaces can be concluded by looking at their *Sobolev numbers* $k + \alpha$ for $C^{k+\alpha}(\Omega)$ and $k - n/p$ for $W^{k,p}(\Omega)$. A continuous embedding $id : X \to Y$ exists if $X \subset Y$ and if the Sobolev number of X is larger than that of Y. In case of equality of the Sobolev numbers, such embeddings may or may not exist.

1.2.2   Fractional Sobolev spaces

For $s \in \mathbb{R}_{>0}$ let $\lfloor s \rfloor := \max\{m \in \mathbb{N} : m \leq s\}$ and define the *Fractional Sobolev spaces* (or *Sobolev–Slobodeckij spaces*) by

$$W^{s,p}(\Omega) := \{w : \Omega \to \mathbb{R} : \|w\|_{W^{s,p}(\Omega)} < \infty\},$$

where, with $m := \lfloor s \rfloor$,

$$\|w\|_{W^{s,p}(\Omega)}^p := \|w\|_{W^{m,p}(\Omega)}^p + [w]_{W^{s,p}(\Omega)}^p,$$

$$[w]_{W^{s,p}(\Omega)}^p := \sum_{|\alpha|=m} \int_\Omega \int_\Omega \frac{|\partial^\alpha w(\boldsymbol{x}) - \partial^\alpha w(\boldsymbol{y})|^p}{|\boldsymbol{x} - \boldsymbol{y}|^{p(s-m)+n}} \, d\boldsymbol{x} \, d\boldsymbol{y}.$$

This is, for $s \geq 0$ and $p \in (1, \infty)$, a separable, reflexive Banach space. In case $p = 2$ it is a Hilbert space that is denoted by H^s. $C^\infty(\overline{\Omega})$ is dense in $W^{s,p}(\Omega)$. $W_0^{s,p}(\Omega)$ is defined to be the closure of $C_0^\infty(\overline{\Omega})$ with respect to $\|\cdot\|_{W^{s,p}(\Omega)}$ and the vector valued function space is defined again componentwise. The dual spaces are denoted by $W^{-s,p'}(\Omega)$ or $H^{-s}(\Omega)$, respectively. These spaces are defined for example in [3, Ch. 14] [11, Ch. B.3.1].

An alternative norm equivalent construction for $p = 2$ is as follows: let E be a continuous extension operator $E_\Omega : f : \Omega \to \mathbb{R} \mapsto E_\Omega f : \mathbb{R}^n \to \mathbb{R}$ (see e. g. [3, Ch. 1.4] [11, Ch. B.3.2]) and let

$$\|f\|_{H^s(\Omega)}^2 := \int_{\mathbb{R}^n} |\widehat{E_\Omega f}(\boldsymbol{\xi})|^2 \left(1 + |\boldsymbol{\xi}|^2\right)^s d\boldsymbol{\xi},$$

where $\hat{}$ denotes the Fourier transform as in Section 1.1.3. Details can be found in [26].

1.2.3   Traces

One can also define Lebesgue and Sobolev spaces on the boundary manifold $\partial\Omega$ as $L^p(\partial\Omega), W^{m,p}(\partial\Omega), \ldots$, or other smooth lower dimensional manifolds in $\overline{\Omega}$. There exists a continuous *trace operator*

$$\gamma^{\partial\Omega} : W^{m,p}(\Omega) \to W^{r,q}(\partial\Omega)$$

such that $\gamma^{\partial\Omega} v = v\big|_{\partial\Omega}$ for all $v \in C^\infty(\overline{\Omega})$, if $m > r$ and $m - n/p \geq r - (n-1)/q$ (see [3, Ch. 1.6], [11, Ch. B.3.5], or [21, Ch. 3.2.1]). Note also, that $\gamma^{\partial\Omega}$ is surjective in this case: for each $g \in W^{r,q}(\partial\Omega)$ there exists $v \in W^{m,p}(\Omega)$ such that $\gamma^{\partial\Omega} v = g$.

1.2.4 The space $H(\mathrm{div}, \Omega)$

The linear space

$$H(\mathrm{div}, \Omega) := \{ \boldsymbol{v} \in L^2(\Omega)^3 \; : \; \boldsymbol{\nabla} \cdot \boldsymbol{v} \in L^2(\Omega) \}$$

with the norm

$$\| \boldsymbol{v} \|^2_{H(\mathrm{div}, \Omega)} := \| \boldsymbol{v} \|^2_{L^2(\Omega)^3} + \| \boldsymbol{\nabla} \cdot \boldsymbol{v} \|^2_{L^2(\Omega)}$$

is a Hilbert space. It has $C^\infty(\overline{\Omega})^3$ as a dense subset. We define the trace operator $\gamma_{\mathrm{n}}^{\partial\Omega}$ by $\gamma_{\mathrm{n}}^{\partial\Omega} \boldsymbol{v} = \boldsymbol{n} \cdot \boldsymbol{v}|_{\partial\Omega}$ for all $\boldsymbol{v} \in C^\infty(\overline{\Omega})^3$. This trace operator has a continuous extension $H(\mathrm{div}, \Omega) \to H^{-1/2}(\partial\Omega)$. We define $H_0(\mathrm{div}, \Omega)$ by the closure of $C_0^\infty(\overline{\Omega})^3$ with respect to $\| \cdot \|_{H(\mathrm{div}, \Omega)}$. In this case it holds

$$\gamma_{\mathrm{n}}^{\partial\Omega} \big(H_0(\mathrm{div}, \Omega) \big) = 0.$$

The integration by parts formula reads: for all $\boldsymbol{v} \in H(\mathrm{div}, \Omega)$ and $w \in H^1(\Omega)$ there holds

$$\int_\Omega \{ \boldsymbol{v} \cdot \boldsymbol{\nabla} w + \boldsymbol{\nabla} \cdot \boldsymbol{v} \, w \} = \langle \gamma_{\mathrm{n}}^{\partial\Omega} \boldsymbol{v}, w \rangle_{H^{-\frac{1}{2}}(\partial\Omega) \times H^{\frac{1}{2}}(\partial\Omega)}.$$

1.2.5 The space $H(\mathrm{curl}, \Omega)$

The linear space

$$H(\mathrm{curl}, \Omega) := \{ \boldsymbol{v} \in L^2(\Omega)^3 \; : \; \boldsymbol{\nabla} \times \boldsymbol{v} \in L^2(\Omega)^3 \}$$

with the norm

$$\| \boldsymbol{v} \|^2_{H(\mathrm{curl}, \Omega)} := \| \boldsymbol{v} \|^2_{L^2(\Omega)^3} + \| \boldsymbol{\nabla} \times \boldsymbol{v} \|^2_{L^2(\Omega)^3}$$

is a Hilbert space. It has $C^\infty(\overline{\Omega})^3$ as a dense subset. We define the trace operator $\gamma_{\mathrm{t}}^{\partial\Omega}$ by $\gamma_{\mathrm{t}}^{\partial\Omega} \boldsymbol{v} = \boldsymbol{n} \times \boldsymbol{v}|_{\partial\Omega}$ for all $\boldsymbol{v} \in C^\infty(\overline{\Omega})^3$. This trace operator has a continuous extension $H(\mathrm{curl}, \Omega) \to H^{-1/2}(\partial\Omega)^3$. We define $H_0(\mathrm{curl}, \Omega)$ by the closure of $C_0^\infty(\overline{\Omega})^3$ with respect to $\| \cdot \|_{H(\mathrm{curl}, \Omega)}$. In this case it holds

$$\gamma_{\mathrm{t}}^{\partial\Omega} \big(H_0(\mathrm{curl}, \Omega) \big) = \boldsymbol{0}.$$

Moreover, we define the trace operator $\gamma_{\mathrm{T}}^{\partial\Omega}$ by $\gamma_{\mathrm{T}}^{\partial\Omega} \boldsymbol{v} = (\boldsymbol{n} \times \boldsymbol{v}) \times \boldsymbol{n}|_{\partial\Omega}$ for all $\boldsymbol{v} \in C^\infty(\overline{\Omega})^3$. This trace operator has also a continuous extension $H(\mathrm{curl}, \Omega) \to H^{-1/2}(\partial\Omega)^3$. More precisely, one finds with $Y := \mathrm{range}(\gamma_{\mathrm{t}}^{\partial\Omega})$ and $Y' = \mathrm{range}(\gamma_{\mathrm{T}}^{\partial\Omega})$ that the integration by parts formula reads: for all $\boldsymbol{v}, \boldsymbol{w} \in H(\mathrm{curl}, \Omega)$ holds

$$\int_\Omega \{ \boldsymbol{\nabla} \times \boldsymbol{v} \cdot \boldsymbol{w} - \boldsymbol{v} \cdot \boldsymbol{\nabla} \times \boldsymbol{w} \} = \langle \gamma_{\mathrm{T}}^{\partial\Omega} \boldsymbol{w}, \gamma_{\mathrm{t}}^{\partial\Omega} \boldsymbol{v} \rangle_{Y' \times Y}$$

[21, Ch. 3.5].

1.2.6   Periodic function spaces

We let $\Gamma \subset \mathbb{R}^d$ be a lattice as in Section 1.1.7. We have seen that periodic functions are determined on a fundamental compact cell Ω. Another viewpoint is to define the equivalence relation $\boldsymbol{x} \sim \boldsymbol{y} \Leftrightarrow \boldsymbol{x} - \boldsymbol{y} \in \Gamma$. The corresponding residue class, denoted by $\mathbb{R}^d/\Gamma$, is a compact manifold and it can be identified with Ω. Thus we let for $m > 0$

$$H_{\mathrm{per}}^m(\Omega) := \left\{ v\big|_\Omega : v \in H^m(\mathbb{R}^3/\Gamma) \right\}.$$

This can be used to define

$$H_{\mathrm{per}}(\mathrm{curl}, \Omega) := \left\{ \boldsymbol{v}\big|_\Omega : \boldsymbol{v} \in H(\mathrm{curl}, \mathbb{R}^3/\Gamma) \right\}$$

and $H_{\mathrm{per}}(\mathrm{div}, \Omega)$ likewise. Note that these are functions that are defined on Ω and for which the corresponding traces are well defined on the identified portions of $\partial\Omega$.

1.2.7   Compact embeddings

For a bounded domain $\Omega \subset \mathbb{R}^3$ with Lipschitz-continuous boundary, the embeddings $H^s(\Omega) \hookrightarrow L^2(\Omega)$ are continuous and *compact* for $s > 0$ (i. e., a bounded sequence in $H^s(\Omega)$ has a subsequence that converges in $L^2(\Omega)$). Especially, $H^1(\Omega)$ is compactly embedded in $L^2(\Omega)$ (Rellich's theorem). Obviously, $H(\mathrm{curl}, \Omega)$ and $H(\mathrm{div}, \Omega)$ are subspaces of $L^2(\Omega)^3$ that both contain $H^1(\Omega)^3$,

$$H^1(\Omega)^3 \subsetneq H(\mathrm{curl}, \Omega), H(\mathrm{div}, \Omega) \subsetneq L^2(\Omega)^3.$$

However, neither $H(\mathrm{curl}, \Omega)$ and $H(\mathrm{div}, \Omega)$ nor $H(\mathrm{curl}, \Omega) \cap H(\mathrm{div}, \Omega)$ are compactly embedded in $L^2(\Omega)^3$ [2, Prop. 2.7]. Taking boundary values into account can change the situation as we see from the following theorem.

Theorem 1.2.1. *Let $\Omega \subset \mathbb{R}^3$ be a bounded domain with Lipschitz-continuous boundary. The space $X_{\mathrm{N}}(\Omega) := H_0(\mathrm{curl}, \Omega) \cap H(\mathrm{div}, \Omega)$ is continuously embedded in $H^s(\Omega)^3$ for some $s > 1/2$ and for all $\boldsymbol{v} \in X_{\mathrm{N}}(\Omega)$ holds*

$$\|\boldsymbol{v}\|_{H^s(\Omega)^3} \leq C\big(\|\boldsymbol{v}\|_{L^2(\Omega)^3} + \|\boldsymbol{\nabla} \times \boldsymbol{v}\|_{L^2(\Omega)^3} + \|\boldsymbol{\nabla} \cdot \boldsymbol{v}\|_{L^2(\Omega)^3} \big),$$

with a constant $C > 0$ that depends on Ω and s. Especially, $X_{\mathrm{N}}(\Omega)$ is compactly embedded in $L^2(\Omega)^3$. If the boundary of Ω is of class $C^{1,1}$ or Ω is convex, then the embedding is valid for $s = 1$.

The same result holds true if we replace $X_{\mathrm{N}}(\Omega)$ by $X_{\mathrm{T}}(\Omega) := H(\mathrm{curl}, \Omega) \cap H_0(\mathrm{div}, \Omega)$.

Proof. See [2].   □

1.2.8 Decompositions

The curl and div operators will also play a role in the decomposition of vector fields: They can be split into a sum of a vector field in the kernel of div and one in the kernel of curl. We start by giving a complete characterisation of the kernel of the curl operator for simply connected domains.

Theorem 1.2.2 (De Rham Theorem). *Let $\Omega \subset \mathbb{R}^3$ be a bounded and simply connected domain with Lipschitz-continuous boundary. Let $\boldsymbol{v} \in H_0(\mathrm{curl}, \Omega)$. Then $\boldsymbol{\nabla} \times \boldsymbol{v} = \boldsymbol{0}$ in Ω is necessary and sufficient for the existence of a scalar potential $q \in H^1(\Omega)$, such that q is constant on each connected component of $\partial\Omega$ and $\boldsymbol{v} = \boldsymbol{\nabla}q$. If $\partial\Omega$ consists of a single component, we can choose $q \in H_0^1(\Omega)$.*

Proof. [21, Thm. 3.37, Thm. 3.41, Thm. 4.3, Rem. 4.4]. □

Theorem 1.2.3 (Helmholtz decomposition). *Let $\Omega \subset \mathbb{R}^3$ be a bounded domain with Lipschitz-continuous boundary. For each $\boldsymbol{v} \in H_0(\mathrm{curl}, \Omega)$, there exists $\boldsymbol{z} \in H_0(\mathrm{curl}, \Omega)$ with $\boldsymbol{\nabla} \cdot \boldsymbol{z} = 0$ (i. e., $\boldsymbol{z} \in X_N(\Omega)$) and $q \in H^1(\Omega)$ such that q is constant on each connected component of $\partial\Omega$ and*

$$\boldsymbol{v} = \boldsymbol{z} + \boldsymbol{\nabla}q.$$

If $\partial\Omega$ consists of a single component, we can choose $q \in H_0^1(\Omega)$. This splitting is orthogonal with respect to the $L^2(\Omega)^3$ and the $H(\mathrm{curl}, \Omega)$ inner product. The mappings $\boldsymbol{v} \mapsto \boldsymbol{z}$ and $\boldsymbol{v} \mapsto \boldsymbol{\nabla}q$ are continuous, that is, it holds

$$\|\boldsymbol{z}\|_{H(\mathrm{curl},\Omega)} + \|\boldsymbol{\nabla}q\|_{L^2(\Omega)^3} \leq \|\boldsymbol{v}\|_{H(\mathrm{curl},\Omega)}.$$

This decomposition of $\boldsymbol{v}$ into $\boldsymbol{z}$ and $\boldsymbol{\nabla}q$ is called Helmholtz decomposition.

Proof. See [21, Lem. 4.5] for $\varepsilon_\mathrm{r} = 1$ and homogeneous boundary conditions. The idea is to define, for given $\boldsymbol{v}$, q as the solution of $-\Delta q = -\boldsymbol{\nabla}\cdot\boldsymbol{v}$ in Ω, $q = 0$ on $\partial\Omega$ (in the single-connected case), in the weak sense: Existence and uniqueness of q follow from the Lax–Milgram Theorem 1.2.5 with $a(q, \xi) := \int_\Omega \boldsymbol{\nabla}q\cdot\boldsymbol{\nabla}\xi$ and $F(\xi) := \int_\Omega \boldsymbol{v} \cdot \boldsymbol{\nabla}\xi$ for $H := H_0^1(\Omega)$ (and with $\alpha = C = 1$). Then let $\boldsymbol{z} := \boldsymbol{v} - \boldsymbol{\nabla}q$. □

The following decomposition, however, yields a more regular part in the first term.

Theorem 1.2.4 (Regular decomposition). *Let $\Omega \subset \mathbb{R}^3$ be a bounded domain with Lipschitz-continuous boundary. For each $\boldsymbol{v} \in X_N(\Omega)$, there exists $\boldsymbol{z} \in H^1(\Omega)^3$ with $\boldsymbol{\nabla}\cdot\boldsymbol{z} = 0$ and $q \in H_0^1(\Omega)$ such that*

$$\boldsymbol{v} = \boldsymbol{z} + \boldsymbol{\nabla}q.$$

The mappings $\boldsymbol{v} \mapsto \boldsymbol{z}$ and $\boldsymbol{v} \mapsto \boldsymbol{\nabla}q$ are continuous, that is, it holds

$$\|\boldsymbol{z}\|_{H^1(\Omega)^3} + \|\boldsymbol{\nabla}q\|_{L^2(\Omega)^3} \leq C\|\boldsymbol{v}\|_{H(\mathrm{curl},\Omega)}.$$

with a constant $C > 0$ depending on Ω. A corresponding result holds for $\boldsymbol{v} \in X_T(\Omega)$ with $q \in H^1(\Omega)$.

Proof. [6, Thm. 3.4].   □

Both theorems can be formulated more general concerning the structure of the decomposition, depending on the topology of Ω, however, we limit our considerations in view of our later applications in Section 2.2.

1.2.9   Two functional analytic tools to prove existence

The following well-known theorem is a consequence of the Riesz representation theorem from functional analysis in case of positive definite forms [21, Thm. 2.17, Thm. 2.21].

Theorem 1.2.5 (Lax–Milgram Theorem). *Assume that the forms a and F satisfy the following properties on a Hilbert space H.*

(1) *$a : H \times H \to \mathbb{R}$ is a continuous and coercive bilinear form, that is,*

$$|a(v, w)| \leq C\|v\|_H \|w\|_H \qquad \text{for all } v, w \in H,$$
$$a(w, w) \geq \alpha\|w\|_H^2 \qquad \text{for all } w \in H,$$

for positive constants C, α.

(2) *$F : H \to \mathbb{R}$ is a continuous linear form.*

Then there is a unique solution $u \in H$ to the problem

$$a(u, v) = F(v) \qquad \text{for all } v \in V$$

that depends continuously on data, i. e., it holds

$$\|u\|_H \leq \frac{1}{\alpha}\|F\|_{H'}.$$

The next result generalises the Lax–Milgram Theorem 1.2.5 to problems with linear constraints [21, Thm. 2.25].

Theorem 1.2.6 (Babuška–Brezzi Theorem). *Assume that the forms a, b and F satisfy the following properties on a pair of Hilbert spaces H, M and*

$$K := \big\{ v \in H \ : \ b(q, v) = 0 \text{ for all } q \in M \big\}.$$

(1) *$a : H \times H \to \mathbb{R}$ is a continuous bilinear form and coercive on the closed subspace K of H, that is,*

$$a(w, w) \geq \alpha\|w\|_H^2 \qquad \text{for all } w \in K$$

for some positive constant α.

(2) $b : M \times H \to \mathbb{R}$ *is a continuous bilinear form and it satisfies the* inf–sup *condition*

$$\inf_{q \in M \setminus \{0\}} \sup_{v \in H \setminus \{0\}} \frac{b(q, v)}{\|q\|_M \|v\|_H} \geq \beta > 0$$

for some constant β.

(3) $F : H \to \mathbb{R}$ *is a continuous linear form.*

Then there is a unique solution $[u, p] \in K \times M$ to the problem

$$a(u, v) + b(p, v) = F(v) \qquad \text{for all } v \in V,$$
$$b(q, u) = 0 \qquad \text{for all } q \in Q,$$

that depends continuously on data, i. e., it holds

$$\|u\|_H \leq \frac{1}{\alpha} \|F\|_{H'} \quad \text{and} \quad \|p\|_M \leq \frac{1}{\beta}\left(1 + \frac{1}{\alpha}\right)\|F\|_{H'}.$$

Bibliography

[1] G. P. Agrawal. *Nonlinear Fiber Optics.* Academic Press, San Diego, 2006.

[2] C. Amrouche, C. Bernardi, M. Dauge, and V. Girault. Vector potentials in three–dimensional nonsmooth domains. *Math. Meth. Appl. Sci.*, 21:823–864, 1998.

[3] S. C. Brenner and L. R. Scott. *The mathematical theory of finite element methods*, volume 15 of *Texts in Applied Mathematics*. Springer, New York, 1994.

[4] D. A. G. Bruggeman. Berechnung verschiedener physikalischer Konstanten von heterogenen Substanzen, I. Dielektrizitätskonstanten und Leitfähigkeiten der Mischkörper aus isotropen Substanzen. *Ann. Phys.*, 24:636664, 1935.

[5] A. Bulovyatov. *Parallel multigrid methods for the band structure computation of 3D photonic crystals with higher order finite elements.* Phd thesis, Karlsruhe Institute of Technology, 2010.

[6] M. Costabel, M. Dauge, and S. Nicaise. Singularities of Maxwell interface problems. *Math. Model. Numer. Anal.*, 33:627–649, 1999.

[7] S. J. Cox and D. C. Dobson. Maximizing band gaps in two-dimensional photonic crystals. *SIAM J. Appl. Math.*, 59:2108–2120, 1999.

[8] P. Donnat and J. Rauch. Global solvability of the Maxwell–Bloch equations from nonlinear optics. *Arch. Rat. Mech. Anal.*, 136:291–303, 1996.

[9] E. Dumas. Global existence for Maxwell–Bloch systems. *J. Differential Equations*, 219:484 509, 2005.

[10] C. Engström and M. Richter. On the spectrum of an operator pencil with applications to wave propagation in periodic and frequency dependent materials. *SIAM J. Appl. Math.*, 70:231 247, 2009.

[11] A. Ern and J.-L. Guermond. *Theory and practice of finite elements*, volume 159 of *Applied Mathematical Sciences*. Springer, New York, 2004.

[12] A. Figotin and P. Kuchment. Band gap structure of spectra of periodic dielectric and acoustic media. I. Scalar model. *SIAM J. Appl. Math.*, 56:68–88, 1996.

[13] A. Figotin and P. Kuchment. Band gap structure of spectra of periodic dielectric and acoustic media. II. Two-dimensional photonic crystals. *SIAM J. Appl. Math.*, 56:1561–1620, 1996.

[14] A. García-Martín, D. Hermann, F. Hagmann, K. Busch, and P. Wölfle. Defect computations in photonic crystals: A solid state approach. *Nanotechnology*, 14:177–183, 2003.

[15] J. C. M. Garnett. Colors in metal glasses and metal films. *Trans. R. Soc.*, 203:385420, 1904.

[16] V. Hoang, M. Plum, and C. Wieners. A computer-assisted proof for photonic band gaps. *Zeitschrift für Angewandte Mathematik und Physik*, 60:1–18, 2009.

[17] J. D. Jackson. *Classical electrodynamics*. John Wiley & Sons, New York, 1975.

[18] J. D. Joannopoulos, S. G. Johnson, J. N. Winn, and R. D. Meade. *Molding the Flow of Light (2. ed.)*. Princeton University Press, Princeton (NJ), 2008.

[19] C. Y. Kao, S. Osher, and E. Yablonovitch. Maximizing band gaps in two-dimensional photonic crystals by using level set methods. *Appl. Phys. B*, 81:235–244, 2005.

[20] T. Kato. *Perturbation theory for linear operators* (2nd Edition). Grundlehren der mathematischen Wissenschaften. Springer, Berlin, 1976.

[21] P. Monk. *Finite Element Methods for Maxwell's Equations*. Clarendon Press, Oxford, 2003.

[22] R. Pantell and H. Puthoff. *Fundamentals of Quantum Electronics*. Wiley, New York, 1969.

[23] J. Pendry. *Fundamentals and Applications of Negative Refraction in Metamaterials*. Princeton University Press, New Jersey, 2007.

[24] M. Richter. *Optimization of photonic bandgaps*. Phd thesis, Karlsruhe Institute of Technology, 2010.

[25] F. Tisseur and K. Meerbergen. The quadratic eigenvalue problem. *SIAM Rev.*, 43:235–286, 2001.

[26] J. Wloka. *Partial differential equations*. Cambridge University Press, Cambridge, 1987.

Chapter 2

Photonic bandstructure calculations

2.1 Approximation of Maxwell eigenvalues and eigenfunctions

by Willy Dörfler

2.1.1 The Maxwell eigenvalue problem

We recall that for the time-harmonic case with frequency ω we derived for the spatially varying part of the electric field $\boldsymbol{x} \mapsto \boldsymbol{E}(\boldsymbol{x})$ the equation (1.17),

$$\boldsymbol{\nabla} \times \left(\frac{1}{\mu} \boldsymbol{\nabla} \times \boldsymbol{E} \right) = \omega^2 \varepsilon \boldsymbol{E}$$

with the additional constraint

$$\boldsymbol{\nabla} \cdot \left(\varepsilon \boldsymbol{E} \right) = 0.$$

For the material parameters we set $\mu = \mu_0 \mu_{\mathrm{r}}$ and $\varepsilon = \varepsilon_0 \varepsilon_{\mathrm{r}}$. Moreover, we assume here that Ω is bounded in $\mathbb{R}^3$ and we impose the boundary condition (1.20), $\boldsymbol{n} \times \boldsymbol{u} = \boldsymbol{0}$, of a perfect conductor on $\partial\Omega$. With the definitions $\lambda := (\omega/c)^2$ (c the speed of light, Section 1.1.4) and $\boldsymbol{u} := \boldsymbol{E}$ the eigenvalue problem now reads

$$\boldsymbol{\nabla} \times \left(\frac{1}{\mu_{\mathrm{r}}} \boldsymbol{\nabla} \times \boldsymbol{u} \right) = \lambda \varepsilon_{\mathrm{r}} \boldsymbol{u} \qquad \text{in } \Omega,$$

$$\boldsymbol{\nabla} \cdot \left(\varepsilon_{\mathrm{r}} \boldsymbol{u} \right) = 0 \qquad \text{in } \Omega,$$

$$\boldsymbol{n} \times \boldsymbol{u} = \boldsymbol{0} \qquad \text{on } \partial\Omega.$$

λ is called Maxwell eigenvalue. Note that the problem for $\boldsymbol{u} := \boldsymbol{H}$ would lead to equations with ε_{r} and μ_{r} interchanged and the boundary condition $\boldsymbol{n} \cdot \boldsymbol{u} = 0$ (1.20). We will continue with the first version and the additional setting $\mu_{\mathrm{r}} = 1$, which is a reasonable choice for many applications.

The eigenvalue problem will be connected to the following problem, the (strong) *curl–curl source problem*: Let a (suitable) vector field $\boldsymbol{f}$ be given. Assume that we can define $\boldsymbol{u} := \boldsymbol{T}\boldsymbol{f}$ as the unique solution to the problem

$$
\begin{aligned}
\boldsymbol{\nabla} \times \boldsymbol{\nabla} \times \boldsymbol{u} &= \varepsilon_{\mathrm{r}} \boldsymbol{f} && \text{in } \Omega, \\
\boldsymbol{\nabla} \cdot (\varepsilon_{\mathrm{r}} \boldsymbol{u}) &= 0 && \text{in } \Omega, \\
\boldsymbol{n} \times \boldsymbol{u} &= \boldsymbol{0} && \text{on } \partial\Omega.
\end{aligned}
\tag{2.1}
$$

This defines a linear operator $\boldsymbol{T}$. The connection to the eigenvalue problem is that we now seek a nontrivial $\boldsymbol{u}$ such that $\boldsymbol{u} = \boldsymbol{T}(\lambda \boldsymbol{u})$ or $\boldsymbol{T}\boldsymbol{u} = \boldsymbol{u}/\lambda$ (for $\lambda \neq 0$). Thus we have to study the spectrum of $\boldsymbol{T}$. The same can be done in the discrete case. This will be the guideline for the following section.

We restrict the considerations to the following classes of domains Ω and coefficients $\varepsilon_{\mathrm{r}}, \mu_{\mathrm{r}}$.

Assumption 2.1.1.

(1) Let $\Omega \subset \mathbb{R}^3$ be a bounded domain with a polygonal simply connected boundary.

(2) Let $\varepsilon_{\mathrm{r}} : \Omega \to \mathbb{R}$ be piecewise constant with two-sided bounds $1 \leq \varepsilon_{\mathrm{r}} \leq \varepsilon_{\infty}$ for some $\varepsilon_{\infty} \geq 1$, $\mu_{\mathrm{r}} = 1$.

For more general conditions on ε_{r} and μ_{r}, such as $\varepsilon_{\mathrm{r}}, \mu_{\mathrm{r}} \in \mathbb{C}$ or $\varepsilon_{\mathrm{r}}, \mu_{\mathrm{r}} \in \mathbb{R}^{3,3}$, and the related different techniques, see [19] [26, Ch. 4.2, 4.6].

2.1.2 The weak curl–curl source problem

The second equation in (2.1) is interpreted as a linear constraint on the first equation. We choose a *Lagrange 'parameter'* p, multiply the first equation by an arbitrary $\boldsymbol{v} \in H_0(\mathrm{curl}, \Omega)$ and the second with arbitrary $q \in H_0^1(\Omega)$ and integrate by parts to get the *weak curl–curl source problem*

$$
\int_\Omega \{\boldsymbol{\nabla} \times \boldsymbol{u} \cdot \boldsymbol{\nabla} \times \boldsymbol{v} + \varepsilon_{\mathrm{r}} \boldsymbol{\nabla} p \cdot \boldsymbol{v}\} = \int_\Omega \varepsilon_{\mathrm{r}} \boldsymbol{f} \cdot \boldsymbol{v} \qquad \text{for all } \boldsymbol{v} \in H_0(\mathrm{curl}, \Omega), \tag{2.2}
$$

$$
\int_\Omega \varepsilon_{\mathrm{r}} \boldsymbol{\nabla} q \cdot \boldsymbol{u} = 0 \qquad \text{for all } q \in H_0^1(\Omega).
$$

In the following we will use the abbreviations

$$
\begin{aligned}
\boldsymbol{V} &:= H_0(\mathrm{curl}, \Omega) \text{ with } \|\boldsymbol{v}\|_{\boldsymbol{V}} := \|\boldsymbol{v}\|_{H(\mathrm{curl},\Omega)}, \\
Q &:= H_0^1(\Omega) \text{ with } \|q\|_Q := \|\boldsymbol{\nabla} q\|_{L^2(\Omega)^3}.
\end{aligned}
$$

Note that due to Assumption 2.1.1.(2), we have $\boldsymbol{f} \in L^2(\Omega)^3$ if and only if $\varepsilon_\mathrm{r} \boldsymbol{f} \in L^2(\Omega)^3$. The above problem is then formalised as to find $\boldsymbol{u} \in \boldsymbol{V}$ and $p \in Q$ with

$$
\begin{aligned}
a(\boldsymbol{u}, \boldsymbol{v}) + b(p, \boldsymbol{v}) &= F(\boldsymbol{v}) && \text{for all } \boldsymbol{v} \in \boldsymbol{V}, && (2.3) \\
b(q, \boldsymbol{u}) &= 0 && \text{for all } q \in Q,
\end{aligned}
$$

with continuous bilinear forms, resp., linear form

$$
a : \boldsymbol{V} \times \boldsymbol{V} \to \mathbb{R}, \quad a(\boldsymbol{w}, \boldsymbol{v}) := \int_\Omega \boldsymbol{\nabla} \times \boldsymbol{w} \cdot \boldsymbol{\nabla} \times \boldsymbol{v}, \tag{2.4}
$$

$$
b : Q \times \boldsymbol{V} \to \mathbb{R}, \quad b(q, \boldsymbol{v}) := \int_\Omega \varepsilon_\mathrm{r} \boldsymbol{\nabla} q \cdot \boldsymbol{v}, \tag{2.5}
$$

$$
F : \boldsymbol{V} \to \mathbb{R}, \quad F(\boldsymbol{v}) := \int_\Omega \varepsilon_\mathrm{r} \boldsymbol{f} \cdot \boldsymbol{v}. \tag{2.6}
$$

Reduction to a single equation

We incorporate the constraint into the space and therefore seek $\boldsymbol{u} \in \boldsymbol{V}_{\varepsilon_\mathrm{r}}$ with

$$
\boldsymbol{V}_{\varepsilon_\mathrm{r}} := \big\{ \boldsymbol{v} \in \boldsymbol{V} \ : \ b(q, \boldsymbol{v}) = 0 \text{ for all } q \in Q \big\}. \tag{2.7}
$$

Then the second equation is automatically satisfied and it remains

$$
a(\boldsymbol{u}, \boldsymbol{v}) = F(\boldsymbol{v}) \qquad \text{for all } \boldsymbol{v} \in \boldsymbol{V}_{\varepsilon_\mathrm{r}}. \tag{2.8}
$$

Existence

([23] [19, Ch. 4] [26, Ch. 3.4]). Existence and uniqueness for these type of equations will be proved via the Babuška–Brezzi Theorem 1.2.6. In order to establish the requirements, Theorem 1.2.1 would be the right tool for the following arguments if ε_r would be constant. But we now need requirements on $\boldsymbol{\nabla}{\cdot}(\varepsilon_\mathrm{r}\boldsymbol{u})$ instead of $\boldsymbol{\nabla}{\cdot}\boldsymbol{u}$. However, the result can be reformulated correspondingly.

Theorem 2.1.2. *Let* $\Omega \subset \mathbb{R}^3$ *and* ε_r *as in Assumption* 2.1.1. *Then the space* $X_\mathrm{N}(\Omega; \varepsilon_\mathrm{r}) := H_0(\mathrm{curl}, \Omega) \cap H(\mathrm{div}(\varepsilon_\mathrm{r} \,.\,), \Omega)$ *is continuously embedded in* $H^s(\Omega)^3$ *for some* $s > 0$, *more precisely, for each of its elements holds*

$$
\|\boldsymbol{v}\|_{H^s(\Omega)^3} \le C \big(\|\boldsymbol{v}\|_{\boldsymbol{V}} + \|\boldsymbol{\nabla}{\cdot}(\varepsilon_\mathrm{r}\boldsymbol{v})\|_{L^2(\Omega)^3} \big)
$$

with $C > 0$, *depending on* $\Omega, \varepsilon_\mathrm{r}, s$. *Especially,* $\boldsymbol{V}_{\varepsilon_\mathrm{r}}$ *is compactly embedded in* $L^2(\Omega)^3$. *If* ε_r *is uniformly Lipschitz-continuous, then one can find* $s > 1/2$. *If, in addition to that,* Ω *is convex, then we can take* $s = 1$. *A corresponding result holds for* $\boldsymbol{v} \in X_\mathrm{T}(\Omega; \varepsilon_\mathrm{r}) := H(\mathrm{curl}, \Omega) \cap H_0(\mathrm{div}(\varepsilon_\mathrm{r} \,.\,), \Omega)$.

Proof. [19, Thm. 4.1, Lem. 4.2, Cor. 4.3]. Let $\boldsymbol{v} \in X_\mathrm{N}(\Omega; \varepsilon_\mathrm{r})$. Application of the Regular decomposition Theorem 1.2.4, that actually holds also under the stated requirement on $\boldsymbol{v}$, establishes $\boldsymbol{v} = \boldsymbol{z} + \boldsymbol{\nabla}q$ for $\boldsymbol{z} \in H^1(\Omega)^3$ and $q \in Q$. Multiplying by ε_r and taking (formally) the divergence yields the boundary value problem $-\boldsymbol{\nabla} \cdot (\varepsilon_r \boldsymbol{\nabla} q) = \boldsymbol{\nabla} \cdot (\varepsilon_r \boldsymbol{z}) - \boldsymbol{\nabla} \cdot (\varepsilon_r \boldsymbol{v})$ in Ω, $q = 0$ on $\partial\Omega$, to be interpreted in the weak sense. While the second term on the right is in $L^2(\Omega)^3$ by assumption, the first term will be in $L^2(\Omega)^3$ only if ε_r is sufficiently regular (e. g., uniformly Lipschitz), but a distribution otherwise. The result follows from the elaborated regularity theory for elliptic equations with Dirichlet boundary condition that results in the bound $\|\boldsymbol{\nabla} q\|_{H^s(\Omega)^3} \le C(\|\boldsymbol{z}\|_{H^1(\Omega)^3} + \|\boldsymbol{\nabla} \cdot (\varepsilon_r \boldsymbol{v})\|_{L^2(\Omega)^3})$ which leads immediately to the stated bound [14]. We note, that for $\boldsymbol{v} \in \boldsymbol{V}_{\varepsilon_r}$ we get the bound $\|\boldsymbol{v}\|_{H^s(\Omega)^3} \le C\|\boldsymbol{v}\|_{\boldsymbol{V}}$ which proves the compactness.

For the case of $X_T(\Omega; \varepsilon_r)$ we have $\boldsymbol{n} \cdot \boldsymbol{\nabla} q = 0$ as a boundary condition and can refer to regularity results for the homogeneous Neumann problem [14].   □

Theorem 2.1.3 (Poincaré–Friedrich–type inequality). *Let $\Omega \subset \mathbb{R}^3$ and ε_r as in Assumption 2.1.1. Then*

$$\|\boldsymbol{v}\|_{L^2(\Omega)^3} \le C_F \|\boldsymbol{\nabla} \times \boldsymbol{v}\|_{L^2(\Omega)^3} \tag{2.9}$$

holds for all $\boldsymbol{v} \in \boldsymbol{V}_{\varepsilon_r}$ with a constant $C_F > 0$.

Proof. The proof is indirect: Assuming that the assertion is wrong, we have a sequence $\{\boldsymbol{v}_n\}_{n\in\mathbb{N}} \subset \boldsymbol{V}_{\varepsilon_r}$ that satisfies $\|\boldsymbol{v}_n\|_{L^2(\Omega)^3} = 1$ and $\|\boldsymbol{\nabla} \times \boldsymbol{v}_n\|_{L^2(\Omega)^3} \to 0$. Using the compact embedding into $L^2(\Omega)^3$, Theorem 2.1.2, we find a limit $\boldsymbol{v}$ of a subsequence that satisfies: $\|\boldsymbol{v}\|_{L^2(\Omega)^3} = 1$ and $\|\boldsymbol{\nabla} \times \boldsymbol{v}\|_{L^2(\Omega)^3} = 0$. By Theorem 1.2.2 we see that $\boldsymbol{v} = \boldsymbol{\nabla} p$ for some $p \in H^1(\Omega)$. Since $\boldsymbol{n} \times \boldsymbol{v} = \boldsymbol{n} \times \boldsymbol{\nabla} p = \boldsymbol{0}$ on $\partial\Omega$, p is constant on $\partial\Omega$ and we can choose $p = 0$ on $\partial\Omega$. Now p fulfills $\boldsymbol{\nabla}\cdot(\varepsilon_r \boldsymbol{\nabla} p) = 0$ (in the weak sense) with homogeneous boundary conditions and thus $p = 0$ and so $\boldsymbol{v} = \boldsymbol{0}$. This however contradicts $\|\boldsymbol{v}\|_{L^2(\Omega)^3} = 1$ [26, Cor. 3.51, Cor. 4.8].   □

Theorem 2.1.4. *Let $\Omega \subset \mathbb{R}^3$ and ε_r as in Assumption 2.1.1. Then, for any given $\boldsymbol{f} \in L^2(\Omega)^3$, there is a unique solution $[\boldsymbol{u}, p] \in \boldsymbol{V}_{\varepsilon_r} \times Q$ to the weak curl–curl source problem (2.2) and it satisfies the a priori bound*

$$\|\boldsymbol{u}\|_{\boldsymbol{V}} + \|p\|_Q \le C\|\boldsymbol{f}\|_{L^2(\Omega)^3}$$

for $C := (C_F^2 + 2)\varepsilon_\infty$.

Proof. We take a, b, F from (2.5)–(2.6) and identify $H := \boldsymbol{V}$, $K := \boldsymbol{V}_{\varepsilon_r}$, and $M := Q$. From Theorem 2.1.3 we obtain the bound

$$\|\boldsymbol{v}\|_{L^2(\Omega)^3} \le C_F \|\boldsymbol{\nabla} \times \boldsymbol{v}\|_{L^2(\Omega)^3} \qquad \text{for all } \boldsymbol{v} \in \boldsymbol{V}_{\varepsilon_r}.$$

Then we have coercivity of the bilinear form a on $\boldsymbol{V}_{\varepsilon_r}$, since then

$$a(\boldsymbol{v}, \boldsymbol{v}) = \|\boldsymbol{\nabla} \times \boldsymbol{v}\|^2_{L^2(\Omega)^3} = \frac{C_F^2}{C_F^2 + 1}\|\boldsymbol{\nabla} \times \boldsymbol{v}\|^2_{L^2(\Omega)^3} + \frac{1}{C_F^2 + 1}\|\boldsymbol{\nabla} \times \boldsymbol{v}\|^2_{L^2(\Omega)^3}$$

$$\ge \frac{1}{C_F^2 + 1}\left(\|\boldsymbol{v}\|^2_{L^2(\Omega)^3} + \|\boldsymbol{\nabla} \times \boldsymbol{v}\|^2_{L^2(\Omega)^3}\right) = \frac{1}{C_F^2 + 1}\|\boldsymbol{v}\|^2_{\boldsymbol{V}}.$$

To prove the inf–sup–condition choose an arbitrary smooth $q \in Q$ and let $\boldsymbol{v}_q := \nabla q \in L^2(\Omega)^3$. With $\nabla \times \nabla q = \boldsymbol{0}$ we see that $\boldsymbol{v}_q \in H(\text{curl}, \Omega)$. Since $q|_{\partial\Omega} = 0$ we have that $\nabla q|_{\partial\Omega}$ is proportional to the normal vector $\boldsymbol{n}$ to $\partial\Omega$ and thus $\boldsymbol{n} \times \nabla q = \boldsymbol{0}$, hence $\boldsymbol{v}_q \in \boldsymbol{V}$. Thus it holds

$$\sup_{\boldsymbol{v} \in \boldsymbol{V} \setminus \{\boldsymbol{0}\}} \frac{b(q, \boldsymbol{v})}{\|\boldsymbol{v}\|_{\boldsymbol{V}}} \geq \frac{b(q, \boldsymbol{v}_q)}{\|\boldsymbol{v}_q\|_{\boldsymbol{V}}} \geq \min_{\boldsymbol{x} \in \Omega}\{\varepsilon_{\mathrm{r}}(\boldsymbol{x})\} \frac{\|\nabla q\|^2_{L^2(\Omega)^3}}{\|\nabla q\|_{L^2(\Omega)^3}} \geq \|q\|_Q$$

and the inf–sup–condition follows readily from this. Note, that the proof relies on the property $\nabla Q \subset \boldsymbol{V}$. The continuity of the right hand side in (2.8) is easily verified by

$$|F(\boldsymbol{v})| = \left| \int_\Omega \varepsilon_{\mathrm{r}} \boldsymbol{f} \cdot \boldsymbol{v} \right| \leq \|\varepsilon_{\mathrm{r}} \boldsymbol{f}\|_{L^2(\Omega)^3} \|\boldsymbol{v}\|_{L^2(\Omega)^3}$$

$$\leq \varepsilon_\infty \|\boldsymbol{f}\|_{L^2(\Omega)^3} \|\boldsymbol{v}\|_{\boldsymbol{V}}$$

for all $\boldsymbol{v} \in \boldsymbol{V}$. Applying Theorem 1.2.6 proves existence and uniqueness. The estimate follows immediately from (2.8) and the given bounds on a and F. $\qquad\square$

Theorem 2.1.5. *Let $\Omega \subset \mathbb{R}^3$ and ε_{r} as in Assumption 2.1.1. Then the mapping $\boldsymbol{T} : L^2(\Omega)^3 \to \boldsymbol{V}_{\varepsilon_{\mathrm{r}}} \subset L^2(\Omega)^3$, $\boldsymbol{f} \mapsto \boldsymbol{T}\boldsymbol{f} := \boldsymbol{u}$, with $\boldsymbol{u}$ from Theorem 2.1.4, is linear and continuous. Moreover, $\boldsymbol{u} \in H^s(\Omega)^3$ for some $s > 0$ and hence $\boldsymbol{T}$ is compact.*

Proof. The first assertion is a direct consequence of the previous Theorem 2.1.4. The regularity of $\boldsymbol{u}$ and the compactness of $\boldsymbol{T}$ follows from Theorem 2.1.2. $\qquad\square$

Remark 2.1.6. In case $\boldsymbol{f} \in \boldsymbol{V}_{\varepsilon_{\mathrm{r}}}$ we find $p = 0$. Indeed, taking $\boldsymbol{v} := \nabla q \in \boldsymbol{V}$ for $q \in Q$ as a test function, we obtain the weak formulation for $-\nabla \cdot (\varepsilon_{\mathrm{r}} \nabla p) = 0$ in Ω, $p = 0$ on $\partial\Omega$, which is uniquely solved by $p = 0$. However, if $\boldsymbol{f} \notin \boldsymbol{V}_{\varepsilon_{\mathrm{r}}}$, then $p \not\equiv 0$ and thus $\boldsymbol{u}$ is not a weak solution of (2.1) but

$$\nabla \times \nabla \times \boldsymbol{u} + \varepsilon_{\mathrm{r}} \nabla p = \varepsilon_{\mathrm{r}} \boldsymbol{f} \qquad \text{in } \Omega,$$
$$\nabla \cdot (\varepsilon_{\mathrm{r}} \boldsymbol{u}) = 0 \qquad \text{in } \Omega, \qquad (2.10)$$
$$\boldsymbol{n} \times \boldsymbol{u} = \boldsymbol{0} \qquad \text{on } \partial\Omega.$$

Thus, (2.10) is the correct definition for $\boldsymbol{T}\boldsymbol{f}$ in the strong form.

Theorem 2.1.7. *Let $\Omega \subset \mathbb{R}^3$ and ε_{r} as in Assumption 2.1.1. Then there exists a discrete set of positive Maxwell eigenvalues with eigenspaces of finite multiplicity.*

Proof. The weak form of the eigenvalue problem is given by (2.2) with $\boldsymbol{f}$ replaced by $\lambda \boldsymbol{u}$. Taking $\boldsymbol{v} = \boldsymbol{u}$ we see that all eigenvalues are a priori non-negative. Note that according to Remark 2.1.6 we will find $p = 0$. Since the weak curl–curl problem is uniquely solvable, there is no zero eigenvalue. Thus all Maxwell eigenvalues are eigenvalues of $\boldsymbol{T}$, since $\boldsymbol{T}\boldsymbol{u} = 1/\lambda \, \boldsymbol{u}$. The assertion for $\boldsymbol{T}$ follows from its compactness (Theorem 2.1.5). $\qquad\square$

2.1.3   The discrete curl–curl source problem

We choose a *Galerkin method* to find a numerical approximation to (2.1) in that we formulate (2.2) in terms of discrete spaces $\boldsymbol{V}^h \subset \boldsymbol{V}$ and $Q_h \subset Q$ with inherited norms. Usually, these spaces are defined to be piecewise polynomial functions on a decomposition of Ω into a collection of "elements" like tetrahedra or hexahedra. The regularity of these functions at the interfaces has to be such that the trace operators are well defined. The parameter h symbolises the maximal diameter of such elements and it will be the quantity that we refer to in the error estimates which study the behaviour for $h \to 0$. We will here not consider the influence of an approximation of Ω ($\Omega = \Omega_h$) and the permittivity ε_{r} ($\varepsilon_{\mathrm{r}} = \varepsilon_{\mathrm{r},h}$). Thus we seek $\boldsymbol{u}^h \in \boldsymbol{V}^h$, $p_h \in Q_h$ such that, with a, b, F as in (2.4)–(2.4)

$$a(\boldsymbol{u}^h, \boldsymbol{v}^h) + b(\boldsymbol{v}^h, p_h) = F(\boldsymbol{v}^h) \qquad \text{for all } \boldsymbol{v}^h \in \boldsymbol{V}^h, \qquad (2.11)$$
$$b(\boldsymbol{u}^h, q_h) = 0 \qquad \text{for all } q_h \in Q_h.$$

Again, as in Section 2.1.2, we can reduce this system to one equation on a constraint space. Let

$$\boldsymbol{V}^h_{\varepsilon_{\mathrm{r}}} := \left\{ \boldsymbol{w}^h \in \boldsymbol{V}^h \ : \ b(q_h, \boldsymbol{w}^h) = 0 \text{ for all } q_h \in Q_h \right\}$$

and seek $\boldsymbol{u}^h \in \boldsymbol{V}^h_{\varepsilon_{\mathrm{r}}}$ such that

$$a(\boldsymbol{u}^h, \boldsymbol{v}^h) = F(\boldsymbol{v}^h) \qquad \text{for all } \boldsymbol{v}^h \in \boldsymbol{V}^h_{\varepsilon_{\mathrm{r}}}.$$

The discrete problem is again of the type (2.3) and existence und uniqueness requires the validity of the conditions in Theorem 1.2.6. By Theorem 2.1.4 the requirements can be stated as follows.

Theorem 2.1.8. *Assume that an estimate like (2.9) holds with constant C_{dF} for all $\boldsymbol{v}^h \in \boldsymbol{V}^h_{\varepsilon_{\mathrm{r}}}$ and that $\nabla Q_h \subset \boldsymbol{V}^h$. For any given $\boldsymbol{f} \in L^2(\Omega)^3$, there is a unique solution $\boldsymbol{u}^h \in \boldsymbol{V}^h_{\varepsilon_{\mathrm{r}}}$ to the discrete curl–curl source problem (2.11) and it satisfies the bound*

$$\|\boldsymbol{u}^h\|_{\boldsymbol{V}} + \|p_h\|_Q \leq C \|\boldsymbol{f}\|_{L^2(\Omega)^3}$$

for $C := (C_{\mathrm{dF}}^2 + 1)\varepsilon_\infty$.

Proof. We recall the proof of Theorem 2.1.4 and note that the requirement (2.9) guarantees coercivity of a on $\boldsymbol{V}^h_{\varepsilon_{\mathrm{r}}}$ and that the requirement $\nabla Q_h \subset \boldsymbol{V}^h$ suffices to prove the inf–sup–condition.   □

Remark 2.1.9. Note that $\boldsymbol{V}^h_{\varepsilon_{\mathrm{r}}} \not\subset \boldsymbol{V}_{\varepsilon_{\mathrm{r}}}$, thus the required inequality cannot be derived from (2.9). It will be formulated as an assumption in Assumption 2.1.13 and later be proved in Lemma 2.1.23.

In analogy to Theorem 2.1.5 we have the following consequence.

Theorem 2.1.10. *The mapping* $\boldsymbol{T}^h : L^2(\Omega)^3 \rightarrow \boldsymbol{V}^h_{\varepsilon_r} \subset \boldsymbol{V}^h \subset H(\mathrm{curl}, \Omega)$, $\boldsymbol{f} \mapsto \boldsymbol{T}^h \boldsymbol{f} := \boldsymbol{u}^h$ *with* $\boldsymbol{u}^h$ *from Theorem 2.1.8 is linear, continuous, and compact.*

Proof. This is a direct consequence of the previous Theorem 2.1.8. The compactness is trivial since $\boldsymbol{V}^h_{\varepsilon_r}$ is finite dimensional. □

2.1.4 Problems of Maxwell eigenvalue computations

(1) Solutions in general Lipschitz domains can have strong singularities: if Φ is a solution to $\Delta\Phi = 0$ in an L–shaped domain Ω we have $\nabla\Phi \in H^s(\Omega)^3$ for some $s \in (\frac{1}{2}, 1)$. Since $\nabla \times \nabla\Phi = 0$ and $\nabla \cdot \nabla\Phi = 0$, $\boldsymbol{u} = \nabla\Phi$ is an example of a singular solution of a curl–curl equation. In case of discontinuous permittivities, however, one may find examples that $\boldsymbol{u} \in H^s(\Omega)^3$ for any given $s > 0$ [14].

(2) The operator curl has an infinite dimensional kernel, namely the space of all gradients (here $\nabla H^1_0(\Omega)$). The relevant eigenfunctions however lie in the orthogonal component to these gradients. This has been used for the coercivity estimate and for the inf–sup condition in the proof of Theorem 2.1.4.

(3) Finite elements space that do not fit to the curl–div structure will in general produce spurious eigenvalues since the gradient space is not well separated from its orthogonal component [2, Sect. 5].

(4) If one uses H^1–conforming spaces, the method will only compute the projection of the singular solution into $H^1(\Omega)^3$ since the ansatz space is only a proper subspace of the solution space: the method seems to converge, the result however is wrong [12] [26, Lem. 3.5.6] [2, Sect. 5].

2.1.5 Convergence of discrete eigenvalues

With notations from (2.4)–(2.6), the *discrete curl–curl eigenvalue problem* reads

$$a(\boldsymbol{u}^h, \boldsymbol{v}^h) + b(\boldsymbol{v}^h, p_h) = \lambda_h \int_\Omega \varepsilon_r \boldsymbol{u}^h \cdot \boldsymbol{v}^h \qquad \text{for all } \boldsymbol{v}^h \in \boldsymbol{V}^h,$$

$$b(\boldsymbol{u}^h, q_h) = 0 \qquad \text{for all } q_h \in Q_h.$$

An abstract framework

With the operator $\boldsymbol{T}^h$ from Section 2.1.3 at hand, the discrete eigenvalue problem can be formulated as $\boldsymbol{T}^h(\lambda_h \boldsymbol{u}^h) = \boldsymbol{u}^h$ or $\boldsymbol{T}^h \boldsymbol{u}^h = 1/\lambda_h \, \boldsymbol{u}^h$ for $\lambda_h \neq 0$. The convergence of the eigenvalues λ_h for h to 0 is connected to the convergence of eigenvalues of $\boldsymbol{T}^h$ towards eigenvalues of $\boldsymbol{T}$. For linear spaces X, Y, $\mathbb{L}(X, Y)$ will denote the space of linear mappings $X \rightarrow Y$ together with the operator norm.

Theorem 2.1.11. *For the uniform convergence of the spectrum of* $\boldsymbol{T}^h$ *towards the spectrum of* $\boldsymbol{T}$ *it is necessary and sufficient that*

$$\lim_{h \to 0} \|\boldsymbol{T} - \boldsymbol{T}^h\|_{\mathbb{L}(L^2(\Omega)^3, L^2(\Omega)^3)} = 0.$$

Proof. [6, p. 1288]   □

Theorem 2.1.12. *Assume that* $\|\boldsymbol{T} - \boldsymbol{T}^h\|_{\boldsymbol{L}(L^2(\Omega)^3, L^2(\Omega)^3)} \leq Ch^t$ *for* $h \leq h_0$ *with some* $h_0, t > 0$. *Let* $\lambda \in \mathbb{C} \setminus \{0\}$ *be an eigenvalue of* $\boldsymbol{T}$ *with multiplicity* $m_j \in \mathbb{N}$ *and eigenspace* $\mathcal{E}(\lambda_j) \subset \boldsymbol{V}_{\varepsilon_r}$. *Then there are exactly* m_j *discrete eigenvalues* $\lambda_{h,j_1} \ldots \lambda_{h,j_m}$ *of* $\boldsymbol{T}^h$ *that converge to* λ_j. *If we define the approximate eigenvalue by*

$$\tilde{\lambda}_{h,j} := \frac{1}{m_j} \sum_{l=1}^{m_j} \lambda_{h,j_l}$$

and the corresponding approximate eigenspace in $\boldsymbol{V}^h_{\varepsilon_r}$ *by*

$$\tilde{\mathcal{E}}^h(\lambda_{h,j}) := \bigoplus_{l=1}^{m_j} \mathcal{E}^h(\lambda_{h,j_l}),$$

then there exist $h_0 > 0$ *such that for all* $h \in (0, h_0)$

$$|\tilde{\lambda}_j - \tilde{\lambda}_{h,j}| \leq Ch^{2t} \quad \text{and} \quad \delta(\mathcal{E}(\lambda_j), \tilde{\mathcal{E}}^h(\lambda_{h,j})) \leq Ch^t,$$

where δ *measures the distance between linear spaces.*

Proof. See e. g. [3], [6, Thm. 1].   □

Convergence of the discrete curl–curl source problem

Following [6] (or [2, Sect. 13]), we now establish the convergence of solutions of the discrete curl–curl source problem (2.11) towards those of (2.2) under sufficient conditions that are formulated in the following assumption. Here, we slightly changed condition (3) and separated it from the regularity requirement (4) for solutions of Maxwell equations.

For convenience we introduce, for $r, s \geq 0$, the space

$$H^{(r,s)}(\text{curl}, \Omega) := \{\boldsymbol{v} \in H^r(\Omega)^3 \ : \ \boldsymbol{\nabla} \times \boldsymbol{v} \in H^s(\Omega)^3\}$$

with the norm, resp. semi-norm,

$$\|\boldsymbol{v}\|_{H^{(r,s)}(\text{curl},\Omega)} := \|\boldsymbol{v}\|_{H^r(\Omega)^3} + \|\boldsymbol{\nabla} \times \boldsymbol{v}\|_{H^s(\Omega)^3},$$

$$[\boldsymbol{v}]_{H^{(r,s)}(\text{curl},\Omega)} := [\boldsymbol{v}]_{H^r(\Omega)^3} + [\boldsymbol{\nabla} \times \boldsymbol{v}]_{H^s(\Omega)^3}.$$

For $r = s$ we let $H^{(r,r)}(\text{curl}, \Omega) = H^r(\text{curl}, \Omega)$.

Assumption 2.1.13 (Convergence requirements).

(1) *Ellipticity on the discrete kernel*: Assume that there exists a constant $C_{\mathrm{dF}} > 0$ such that

$$\|\boldsymbol{v}^h\|_{L^2(\Omega)^3} \le C_{\mathrm{dF}} \|\nabla \times \boldsymbol{v}^h\|_{L^2(\Omega)^3} \tag{2.12}$$

holds for all $\boldsymbol{v}^h \in \boldsymbol{V}^h_{\varepsilon_{\mathrm{r}}}$.

(2) *Weak approximability*: For some $s > 0$ holds that for all $\boldsymbol{v}^h \in \boldsymbol{V}^h_{\varepsilon_{\mathrm{r}}}$ there exists $\boldsymbol{v} \in \boldsymbol{V}_{\varepsilon_{\mathrm{r}}}$ such that

$$\|\boldsymbol{v}^h - \boldsymbol{v}\|_{L^2(\Omega)^3} \le Ch^s \|\nabla \times \boldsymbol{v}^h\|_{L^2(\Omega)^3}. \tag{2.13}$$

(3) *Strong approximability*: For some $r, r' > 0$ holds that for all suitably regular $\boldsymbol{v} \in \boldsymbol{V}$, there exists $\boldsymbol{v}^h \in \boldsymbol{V}^h$ such that

$$\|\boldsymbol{v} - \boldsymbol{v}^h\|_{\boldsymbol{V}} \le Ch^r \begin{cases} \|\boldsymbol{v}\|_{H^{(r,r')}(\mathrm{curl},\Omega)} & \text{for } r, r' > 0, \\ \|\boldsymbol{v}\|_{H^r(\Omega)^3} & \text{for } r > 1. \end{cases} \tag{2.14}$$

(4) *Regularity*: For the solution $\boldsymbol{u} = \boldsymbol{T}\boldsymbol{f} \in \boldsymbol{V}_{\varepsilon_{\mathrm{r}}}$ of the weak curl–curl problem (2.2) holds one of the a priori bounds

$$\|\boldsymbol{u}\|_{H^{(r,r')}(\mathrm{curl},\Omega)} \le C\|\boldsymbol{f}\|_{L^2(\Omega)^3} \quad \text{for } r, r' > 0$$

$$\text{or} \quad \|\boldsymbol{u}\|_{H^r(\Omega)^3} \le C\|\boldsymbol{f}\|_{L^2(\Omega)^3} \quad \text{for } r > 1.$$

Remark 2.1.14. The interpolation estimate (2.14) can actually be obtained with $\|\boldsymbol{v}\|_{H^r(\Omega)^3} + \|\nabla \times \boldsymbol{v}\|_{L^p(\Omega)}$, for $p > 2$, on the right hand side [6, p. 1286]. For $r > 1$, this last term can be estimated by $\|\boldsymbol{v}\|_{H^r(\Omega)^3}$ using Sobolev embedding results, Section 1.1.4.

The basic error estimate can now be formulated as follows, a result often called Strang lemma in finite element theory [17, Ch. 2.3.2], see also [3].

Lemma 2.1.15. *Using Assumption* 2.1.13. (1) *we have the error bound*

$$\|\boldsymbol{u} - \boldsymbol{u}^h\|_{\boldsymbol{V}} \le C \left(\inf_{\boldsymbol{v}^h \in \boldsymbol{V}^h} \|\boldsymbol{u} - \boldsymbol{v}^h\|_{\boldsymbol{V}} + \sup_{\boldsymbol{w}^h \in \boldsymbol{V}^h_{\varepsilon_{\mathrm{r}}}} \inf_{\boldsymbol{w} \in \boldsymbol{V}_{\varepsilon_{\mathrm{r}}}} \frac{|b(p, \boldsymbol{w}^h - \boldsymbol{w})|}{\|\nabla \times \boldsymbol{w}^h\|_{L^2(\Omega)^3}} \right)$$

between the solutions of the continuous and discrete curl–curl source problem, $\boldsymbol{u} = \boldsymbol{T}\boldsymbol{f} \in \boldsymbol{V}_{\varepsilon_{\mathrm{r}}}$ and $\boldsymbol{u}^h = \boldsymbol{T}^h\boldsymbol{f} \in \boldsymbol{V}^h_{\varepsilon_{\mathrm{r}}}$ (for $\boldsymbol{f} \in L^2(\Omega)^3$), respectively. The constant C can be chosen as $(C_{\mathrm{dF}} + 2)(1 + \sqrt{\varepsilon_\infty})$.

Proof. Let $[\boldsymbol{u}, p] \in \boldsymbol{V}_{\varepsilon_{\mathrm{r}}} \times Q$ and $[\boldsymbol{u}^h, p_h] \in \boldsymbol{V}^h_{\varepsilon_{\mathrm{r}}} \times Q_h$ solve (2.2) and (2.11), respectively. We start to bound $\|\boldsymbol{v}^h - \boldsymbol{u}^h\|_{\boldsymbol{V}}$ for arbitrary $\boldsymbol{v}^h \in \boldsymbol{V}^h_{\varepsilon_{\mathrm{r}}}$ and get

$$\|\boldsymbol{\nabla} \times (\boldsymbol{v}^h - \boldsymbol{u}^h)\|^2_{L^2(\Omega)^3} = \int_\Omega \boldsymbol{\nabla} \times (\boldsymbol{v}^h - \boldsymbol{u}^h) \cdot \boldsymbol{\nabla} \times (\boldsymbol{v}^h - \boldsymbol{u}^h)$$

$$= \int_\Omega \boldsymbol{\nabla} \times (\boldsymbol{v}^h - \boldsymbol{u} + \boldsymbol{u} - \boldsymbol{u}^h) \cdot \boldsymbol{\nabla} \times (\boldsymbol{v}^h - \boldsymbol{u}^h)$$

$$\leq \|\boldsymbol{\nabla} \times (\boldsymbol{v}^h - \boldsymbol{u})\|_{L^2(\Omega)^3} \|\boldsymbol{\nabla} \times (\boldsymbol{v}^h - \boldsymbol{u}^h)\|_{L^2(\Omega)^3} + \left|b(p - p_h, \boldsymbol{v}^h - \boldsymbol{u}^h)\right|,$$

hence with $b(p - p_h, \boldsymbol{v}^h - \boldsymbol{u}^h) = b(p, \boldsymbol{v}^h - \boldsymbol{u}^h - \boldsymbol{w})$ for arbitrary $\boldsymbol{w} \in \boldsymbol{V}_{\varepsilon_{\mathrm{r}}}$

$$\left|b(p - p_h, \boldsymbol{v}^h - \boldsymbol{u}^h)\right| \leq \inf_{\boldsymbol{w} \in \boldsymbol{V}_{\varepsilon_{\mathrm{r}}}} \left|b(p, \boldsymbol{v}^h - \boldsymbol{u}^h - \boldsymbol{w})\right|$$

$$\leq \sup_{\boldsymbol{w}^h \in \boldsymbol{V}^h_{\varepsilon_{\mathrm{r}}}} \inf_{\boldsymbol{w} \in \boldsymbol{V}_{\varepsilon_{\mathrm{r}}}} \frac{|b(p, \boldsymbol{w}^h - \boldsymbol{w})|}{\|\boldsymbol{\nabla} \times \boldsymbol{w}^h\|_{L^2(\Omega)^3}} \|\boldsymbol{\nabla} \times (\boldsymbol{v}^h - \boldsymbol{u}^h)\|_{L^2(\Omega)^3}.$$

This gives

$$\|\boldsymbol{\nabla} \times (\boldsymbol{v}^h - \boldsymbol{u}^h)\|_{L^2(\Omega)^3} \leq \|\boldsymbol{\nabla} \times (\boldsymbol{v}^h - \boldsymbol{u})\|_{L^2(\Omega)^3} + \sup_{\boldsymbol{w}^h \in \boldsymbol{V}^h_{\varepsilon_{\mathrm{r}}}} \inf_{\boldsymbol{w} \in \boldsymbol{V}_{\varepsilon_{\mathrm{r}}}} \frac{|b(p, \boldsymbol{w}^h - \boldsymbol{w})|}{\|\boldsymbol{\nabla} \times \boldsymbol{w}^h\|_{L^2(\Omega)^3}}.$$

The assertion with $\inf_{\boldsymbol{v}^h \in \boldsymbol{V}^h_{\varepsilon_{\mathrm{r}}}}$ follows from

$$\|\boldsymbol{u} - \boldsymbol{u}^h\|_{\boldsymbol{V}} \leq \|\boldsymbol{u} - \boldsymbol{v}^h\|_{\boldsymbol{V}} + \|\boldsymbol{v}^h - \boldsymbol{u}^h\|_{\boldsymbol{V}}$$

$$\leq \|\boldsymbol{u} - \boldsymbol{v}^h\|_{\boldsymbol{V}} + (C_{\mathrm{dF}} + 1)\|\boldsymbol{\nabla} \times (\boldsymbol{v}^h - \boldsymbol{u}^h)\|_{L^2(\Omega)^3}$$

$$\leq (C_{\mathrm{dF}} + 2)\|\boldsymbol{u} - \boldsymbol{v}^h\|_{\boldsymbol{V}} + (C_{\mathrm{dF}} + 1) \sup_{\boldsymbol{w}^h \in \boldsymbol{V}^h_{\varepsilon_{\mathrm{r}}}} \inf_{\boldsymbol{w} \in \boldsymbol{V}_{\varepsilon_{\mathrm{r}}}} \frac{|b(p, \boldsymbol{w}^h - \boldsymbol{w})|}{\|\boldsymbol{\nabla} \times \boldsymbol{w}^h\|_{L^2(\Omega)^3}}.$$

Now let $\boldsymbol{v}^h \in \boldsymbol{V}^h$ and define $\boldsymbol{z_h} = \boldsymbol{\nabla}\phi_h$ by $\int_\Omega \varepsilon_r \boldsymbol{\nabla}\phi_h \cdot \boldsymbol{\nabla} q_h = \int_\Omega \varepsilon_r(\boldsymbol{u} - \boldsymbol{v}^h) \cdot \boldsymbol{\nabla} q_h$ for all $q_h \in Q_h$. $\boldsymbol{z}^h$ uniquely exists with $\|\boldsymbol{z}^h\|_{L^2(\Omega)^3} \leq \sqrt{\varepsilon_\infty}\|\boldsymbol{u} - \boldsymbol{v}^h\|_{L^2(\Omega)^3}$. But then $\boldsymbol{v}^h + \boldsymbol{z}^h \in \boldsymbol{V}^h_{\varepsilon_{\mathrm{r}}}$ and

$$\|\boldsymbol{u} - (\boldsymbol{v}^h + \boldsymbol{z}^h)\|_{\boldsymbol{V}} \leq \|\boldsymbol{u} - \boldsymbol{v}^h\|_{\boldsymbol{V}} + \|\boldsymbol{z}^h\|_{L^2(\Omega)^3} \leq (1 + \sqrt{\varepsilon_\infty})\|\boldsymbol{u} - \boldsymbol{v}^h\|_{\boldsymbol{V}}.$$

This shows the assertion with $\inf_{\boldsymbol{v}^h \in \boldsymbol{V}^h}$. $\square$

Now we can prove an error estimation under the condition that $\boldsymbol{u}$ is sufficiently regular as stated in Assumptions 2.1.13.(4) (cf. [6]).

Theorem 2.1.16. *Under Assumptions 2.1.13 we have* $\boldsymbol{T}^h \to \boldsymbol{T}$ *for* $h \to 0$ *in* $\mathbb{L}(L^2(\Omega)^3, L^2(\Omega)^3)$, *and more precisely, with* $t := \min\{r, s\}$,

$$\|\boldsymbol{T}\boldsymbol{f} - \boldsymbol{T}^h\boldsymbol{f}\|_{\boldsymbol{V}} \leq Ch^t\|\boldsymbol{f}\|_{L^2(\Omega)^3}.$$

Thus the assertions of Theorem 2.1.12 hold true and the convergence of the eigenvalues and eigenspaces can be deduced accordingly.

Proof. We proceed with the result from Lemma 2.1.15 in order to establish an estimate for $\|\boldsymbol{u} - \boldsymbol{u}^h\|_{\boldsymbol{V}}$ in terms of $\|\boldsymbol{f}\|_{L^2(\Omega)^3}$.

Given $\boldsymbol{w}^h$, we see from Assumption 2.1.13.(2) that we can find $\boldsymbol{w} \in \boldsymbol{V}_{\varepsilon_{\mathrm{r}}}$ such that

$$\frac{|b(p, \boldsymbol{w}^h - \boldsymbol{w})|}{\|\boldsymbol{\nabla} \times \boldsymbol{w}^h\|_{L^2(\Omega)^3}} \le \varepsilon_\infty \frac{\|\boldsymbol{w}^h - \boldsymbol{w}\|_{L^2(\Omega)^3}}{\|\boldsymbol{\nabla} \times \boldsymbol{w}^h\|_{L^2(\Omega)^3}} \|p\|_Q \le Ch^s \|\boldsymbol{f}\|_{L^2(\Omega)^3}$$

for some $s > 0$. Thanks to Assumption 2.1.13.(3-4) we can choose $\boldsymbol{v}^h$ so that

$$\|\boldsymbol{u} - \boldsymbol{v}^h\|_{\boldsymbol{V}} \le Ch^r \|\boldsymbol{f}\|_{L^2(\Omega)^3}, \tag{2.15}$$

where one of the cases $r \in (0, 1]$ or $r > 1$ is used. This proves

$$\|\boldsymbol{u} - \boldsymbol{u}^h\|_{\boldsymbol{V}} \le Ch^{\min\{r,s\}} \|\boldsymbol{f}\|_{L^2(\Omega)^3}. \qquad \square$$

Verification of Assumptions 2.1.13

We start with a modification of the Helmholtz decomposition Theorem 1.2.3 for nonconstant ε_{r} and a discrete version of it.

Theorem 2.1.17. *Let $\Omega \subset \mathbb{R}^3$ and ε_{r} as in Assumption 2.1.1. For each $\boldsymbol{v} \in \boldsymbol{V}$, there exists $\boldsymbol{z} \in \boldsymbol{V}_{\varepsilon_{\mathrm{r}}}$ and $q \in H_0^1(\Omega)$ such that*

$$\boldsymbol{v} = \boldsymbol{z} + \boldsymbol{\nabla} q.$$

This splitting is orthogonal with respect to the ε_{r}-weighted $L^2(\Omega)^3$ and the $H(\mathrm{curl}, \Omega)$ inner product. The mappings $\boldsymbol{v} \mapsto \boldsymbol{z}$ and $\boldsymbol{v} \mapsto \boldsymbol{\nabla} q$ are continuous, that is, it holds

$$\|\boldsymbol{z}\|_{\boldsymbol{V}} + \|\boldsymbol{\nabla} q\|_{L^2(\Omega)^3} \le \|\boldsymbol{v}\|_{\boldsymbol{V}}.$$

Proof. See Theorem 1.2.3 or [26, Lem. 4.5]. $\qquad \square$

Corollary 2.1.18. *Let $\Omega \subset \mathbb{R}^3$ and ε_{r} as in Assumption 2.1.1. For each $\boldsymbol{v} \in \boldsymbol{V}$, there exists a unique $\boldsymbol{w} \in \boldsymbol{V}_{\varepsilon_{\mathrm{r}}}$ such that*

$$\boldsymbol{\nabla} \times \boldsymbol{w} = \boldsymbol{\nabla} \times \boldsymbol{v} \quad \text{in } \Omega.$$

Moreover, it holds $\boldsymbol{w} \in H^s(\Omega)^3$ for some $s > 0$ with the bound

$$\|\boldsymbol{w}\|_{H^s(\Omega)^3} \le C \|\boldsymbol{v}\|_{\boldsymbol{V}}$$

for some constant $C > 0$. If ε_{r} is uniformly Lipschitz-continuous, then $s > 1/2$. If, in addition to that, Ω is convex, then we can take $s = 1$. The mapping $\boldsymbol{v} \mapsto \boldsymbol{w} := H_{\varepsilon_{\mathrm{r}}} \boldsymbol{v}$ is also called Hodge mapping [19, Ch. 4.2].

Proof. We apply Theorem 2.1.17 to $\boldsymbol{v} \in \boldsymbol{V}$ and define $\boldsymbol{w} := \boldsymbol{z} \in \boldsymbol{V}_{\varepsilon_{\mathrm{r}}}$. The first assertion then follows immediately. The second assertion then is verified using Theorem 2.1.2.   □

Theorem 2.1.19. *Let $\Omega \subset \mathbb{R}^3$ and ε_{r} as in Assumption 2.1.1. Let $\boldsymbol{V}^h$ and Q_h as in Section 2.1.3 with $\boldsymbol{\nabla} Q_h \subset \boldsymbol{V}^h$. Then, for each $\boldsymbol{v}^h \in \boldsymbol{V}^h$, there exists $\boldsymbol{z}^h \in \boldsymbol{V}^h_{\varepsilon_{\mathrm{r}}}$ and $q_h \in Q_h$ such that*

$$\boldsymbol{v}^h = \boldsymbol{z}^h + \boldsymbol{\nabla} q_h.$$

This splitting is orthogonal with respect to the ε_{r}-weighted $L^2(\Omega)^3$ and the $H(\mathrm{curl}, \Omega)$ inner product. The mappings $\boldsymbol{v}^h \mapsto \boldsymbol{z}^h$ and $\boldsymbol{v}^h \mapsto \boldsymbol{\nabla} q_h$ are continuous, that is, it holds

$$\|\boldsymbol{z}^h\|_{H(\mathrm{curl}, \Omega)} + \|\boldsymbol{\nabla} q_h\|_{L^2(\Omega)^3} \leq \|\boldsymbol{v}^h\|_{H(\mathrm{curl}, \Omega)}.$$

Proof. See Theorem 1.2.3 for the proof or [26, Lem. 4.5].   □

To proceed, we first need some more specific assumptions on the discrete spaces formulated below. It is assumed that our finite element space $\boldsymbol{V}^h$ is constructed on a collection of elements $\mathcal{T}$ of tetrahedra or hexahedra denoted by T with the usual requirements [8, Ch. 3], [17, Ch. 1.2]. In our case we can require $\overline{\Omega} = \cup_{T \in \mathcal{T}} T$. For $T \in \mathcal{T}$ we let $h_T := \mathrm{diam}(T)$ be the diameter of an element T and ω_T be the local domain composed of T and all its neighbouring elements, $\omega_T := \cup_{T' \in \mathcal{T}: T' \cap T \neq \emptyset} \{T'\}$. Note that $h = \max_{T \in \mathcal{T}} h_T$. It is assumed that the mesh is quasi-uniform, meaning that $h_T / h_{T'} \leq C$ for all $T' \in \omega_T$ and all $T \in \mathcal{T}$ with a universal constant $C > 0$.

Assumption 2.1.20. Assume that for $\boldsymbol{V}^h$ there exists a linear projection $\boldsymbol{\Pi}^h : \boldsymbol{V} \cap H^{s_0}(\Omega)^3 \to \boldsymbol{V}^h$, for some $s_0 \in (0, 1]$, such that

(1) $\boldsymbol{\Pi}^h \boldsymbol{v}^h = \boldsymbol{v}^h$ for all $\boldsymbol{v}^h \in \boldsymbol{V}^h$,

(2) $\boldsymbol{\nabla} \times \boldsymbol{v} = \boldsymbol{0}$ implies $\boldsymbol{\Pi}^h \boldsymbol{v} \in \boldsymbol{\nabla} Q_h$ for all $\boldsymbol{v} \in \boldsymbol{V}$,

(3)   For all sufficiently regular $\boldsymbol{v}$ holds

$$\|\boldsymbol{v} - \boldsymbol{\Pi}^h \boldsymbol{v}\|_{L^2(T)^3} \leq C h_T^r \begin{cases} \llbracket \boldsymbol{v} \rrbracket_{H^{(r,r')}(\mathrm{curl}, \omega_T)} & \text{for } r \in [s_0, 1], \, r' > 0, \\ \llbracket \boldsymbol{v} \rrbracket_{H^r(\omega_T)^3} & \text{for } r > 1, \end{cases}$$

$$\|\boldsymbol{\nabla} \times (\boldsymbol{v} - \boldsymbol{\Pi}^h \boldsymbol{v})\|_{L^2(T)^3} \leq C h_T^r \llbracket \boldsymbol{\nabla} \times \boldsymbol{v} \rrbracket_{H^r(\omega_T)^3} \quad \text{for } r \geq s_0$$

(cf. Section 1.2.2). Moreover, if $\boldsymbol{\nabla} \times \boldsymbol{v}$ is piecewise polynomial on $\mathcal{T}$, then

$$\|\boldsymbol{v} - \boldsymbol{\Pi}^h \boldsymbol{v}\|_{L^2(T)^3} \leq C \left(h_T^s \llbracket \boldsymbol{v} \rrbracket_{H^s(\omega_T)^3} + h_T \|\boldsymbol{\nabla} \times \boldsymbol{v}\|_{L^2(\omega_T)^3} \right) \tag{2.16}$$

for some $s \in [s_0, 1]$.

Lemma 2.1.21 (Strong approximability). *Let $\boldsymbol{V}^h$ satisfy Assumption 2.1.20. Then the strong approximability 2.1.13.(3) is valid for $r \geq s_0$.*

Proof. Use Assumption 2.1.20.(3) and sum its square up over $T \in \mathcal{T}$. $\qquad\square$

Lemma 2.1.22 (Weak approximability). *Let $\boldsymbol{V}^h$ satisfy Assumption 2.1.20. Then the weak approximability 2.1.13.(2) is valid for $s \geq s_0$.*

Proof. (Cf. [19, Lem. 4.5]) Let $\boldsymbol{v}^h \in \boldsymbol{V}_{\varepsilon_r}^h$ be given. Define $\boldsymbol{v} \in \boldsymbol{V}_{\varepsilon_r}$ by $\boldsymbol{v} = H_{\varepsilon_r} \boldsymbol{v}^h \in H^s(\Omega)^3$ as in Corollary 2.1.18 for some $s \geq s_0$. By Assumption 2.1.20.(1) we have $\boldsymbol{v}^h - \boldsymbol{\Pi}^h \boldsymbol{v} = \boldsymbol{\Pi}^h(\boldsymbol{v}^h - \boldsymbol{v})$ and by definition $\boldsymbol{\nabla} \times (\boldsymbol{v}^h - \boldsymbol{v}) = \boldsymbol{0}$ which implies that $\boldsymbol{v}^h - \boldsymbol{\Pi}^h \boldsymbol{v} = \boldsymbol{\nabla} q_h$ for some $q_h \in Q_h$ by Assumption 2.1.20.(2). Thus we can derive the bound (note $Q_h \subset Q$)

$$\|\sqrt{\varepsilon_r}(\boldsymbol{v}^h - \boldsymbol{v})\|_{L^2(\Omega)^3}^2 = \int_\Omega \varepsilon_r(\boldsymbol{v}^h - \boldsymbol{v}) \cdot (\boldsymbol{v}^h - \boldsymbol{\Pi}^h \boldsymbol{v} + \boldsymbol{\Pi}^h \boldsymbol{v} - \boldsymbol{v})$$

$$= \int_\Omega \varepsilon_r(\boldsymbol{v}^h - \boldsymbol{v}) \cdot (\boldsymbol{\Pi}^h \boldsymbol{v} - \boldsymbol{v})$$

$$\leq \sqrt{\varepsilon_\infty} \|\sqrt{\varepsilon_r}(\boldsymbol{v}^h - \boldsymbol{v})\|_{L^2(\Omega)^3} \|\boldsymbol{\Pi}^h \boldsymbol{v} - \boldsymbol{v}\|_{L^2(\Omega)^3}$$

which gives

$$\|\boldsymbol{v}^h - \boldsymbol{v}\|_{L^2(\Omega)^3} \leq \|\sqrt{\varepsilon_r}(\boldsymbol{v}^h - \boldsymbol{v})\|_{L^2(\Omega)^3}^2 \leq \sqrt{\varepsilon_\infty} \|\boldsymbol{\Pi}^h \boldsymbol{v} - \boldsymbol{v}\|_{L^2(\Omega)^3}.$$

Now observe that $\boldsymbol{\nabla} \times \boldsymbol{v} = \boldsymbol{\nabla} \times \boldsymbol{v}^h$ is piecewise polynomial so that for some $s \geq s_0$

$$\|\boldsymbol{v}^h - \boldsymbol{v}\|_{L^2(\Omega)^3} \leq Ch^s\big([\![\boldsymbol{v}]\!]_{H^s(\Omega)^3} + \|\boldsymbol{\nabla} \times \boldsymbol{v}\|_{L^2(\Omega)^3}\big) \leq Ch^s \|\boldsymbol{\nabla} \times \boldsymbol{v}^h\|_{L^2(\Omega)^3},$$

where we have used Assumption 2.1.20.(3) ((2.16) squared and summed up over $T \in \mathcal{T}$) and the bound $[\![\boldsymbol{v}]\!]_{H^s(\Omega)^3} \leq C \|\boldsymbol{\nabla} \times \boldsymbol{v}\|_{L^2(\Omega)^3}$ resulting from Theorem 2.1.2 and Theorem 2.1.3. This gives the stated result. $\qquad\square$

Lemma 2.1.23 (Ellipticity on the discrete kernel). *Let $\boldsymbol{V}^h$ satisfy Assumption 2.1.20. Then the ellipticity on the discrete kernel 2.1.13.(1) is valid.*

Proof. (Cf. [19, Lem. 4.7]) Let $\boldsymbol{v}^h \in \boldsymbol{V}_{\varepsilon_r}^h$. According to Theorem 2.1.17 and Theorem 2.1.19 there exists $\boldsymbol{z} \in \boldsymbol{V}_1$, $q \in Q$ and $\boldsymbol{z}^h \in \boldsymbol{V}_1^h$ (that is, with permittivity 1!), $q_h \in Q_h$ such that $\boldsymbol{v}^h = \boldsymbol{z} + \boldsymbol{\nabla} q$ and $\boldsymbol{v}^h = \boldsymbol{z}^h + \boldsymbol{\nabla} q_h$. With this we get the estimate

$$\|\sqrt{\varepsilon_r}\boldsymbol{v}^h\|_{L^2(\Omega)^3}^2 = \int_\Omega \varepsilon_r \boldsymbol{v}^h \cdot \boldsymbol{v}^h = \int_\Omega \varepsilon_r \boldsymbol{v}^h \cdot \boldsymbol{z}^h$$

$$\leq \sqrt{\varepsilon_\infty} \|\sqrt{\varepsilon_r}\boldsymbol{v}^h\|_{L^2(\Omega)^3}\big(\|\boldsymbol{z}\|_{L^2(\Omega)^3} + \|\boldsymbol{z} - \boldsymbol{z}^h\|_{L^2(\Omega)^3}\big).$$

The estimate $\|\boldsymbol{z}\|_{L^2(\Omega)^3} \leq C_F \|\boldsymbol{\nabla} \times \boldsymbol{z}\|_{L^2(\Omega)^3}$ follows from Theorem 2.1.3. Observe that $\boldsymbol{\nabla} \times \boldsymbol{z} = \boldsymbol{\nabla} \times \boldsymbol{z}^h = \boldsymbol{\nabla} \times \boldsymbol{v}^h$ by construction, so that actually it holds that $\boldsymbol{z} = H_{\varepsilon_r} \boldsymbol{z}^h$. From the proof of the weak approximability, Lemma 2.1.22, we thus get $\|\boldsymbol{z} - \boldsymbol{z}^h\|_{L^2(\Omega)^3} \leq Ch^s \|\boldsymbol{\nabla} \times \boldsymbol{z}^h\|_{L^2(\Omega)^3}$ for some $s > 0$. By insertion of these equalities and inequalities we end up with the stated result. $\qquad\square$

Theorem 2.1.24 (Regularity of the electric field). *Let $\boldsymbol{f} \in L^2(\Omega)^3$. Then for the solution $\boldsymbol{E} = \boldsymbol{T}\boldsymbol{f} \in \boldsymbol{V}_{\varepsilon_r}$ of the weak curl–curl problem (2.2) holds the priori bound*

$$\|\boldsymbol{E}\|_{H^{(r',r'')}(\mathrm{curl},\Omega)} \leq C\|\boldsymbol{f}\|_{L^2(\Omega)^3} \quad \text{for some } r' > 0,\, r'' > 1/2.$$

Proof. (Cf. [26, Lem. 7.7]) We first note that the estimate for $\|\boldsymbol{E}\|_{H^{r'}(\Omega)^3}$, for some $r' > 0$, is a consequence of Theorem 2.1.2. Thus it remains to prove a corresponding estimate for $\boldsymbol{w} := \nabla \times \boldsymbol{E}$. For arbitrary $\boldsymbol{f}$, the solution $\boldsymbol{E}$ of (2.2) satisfies $\nabla \times \boldsymbol{w} = \nabla \times \nabla \times \boldsymbol{E} = \varepsilon_r(\boldsymbol{f} - \nabla p)$ and this right hand side is in $L^2(\Omega)^3$ by assumption and Theorem 2.1.4. Hence $\boldsymbol{w} \in H(\mathrm{curl}, \Omega)$ and furthermore $\nabla \cdot \boldsymbol{w} = 0$. Therefore, the trace $\boldsymbol{n} \cdot \boldsymbol{w}$ is well-defined in $H^{-1/2}(\partial\Omega)$ (Section 1.2.4). We claim $\boldsymbol{n} \cdot \boldsymbol{w} = 0$. This would prove $\boldsymbol{w} \in X_{\mathrm{T}}$ and by Theorem 1.2.1 that $\boldsymbol{w} \in H^{r''}(\Omega)^3$ for some $r'' > 1/2$ (note that ε_r does not appear in this argument).

Fix $\boldsymbol{x}_0$ so that Ω is smooth near $\boldsymbol{x}_0$. We may represent $\boldsymbol{E}$ as $\boldsymbol{E} = \alpha\boldsymbol{n} + \beta\boldsymbol{t}$, where $\boldsymbol{n}, \boldsymbol{t}$ are extended (into Ω) normal and tangential vector fields with respect to $\partial\Omega$ and α, β some scalar coefficient functions. $\boldsymbol{n}$ is assumed to be a gradient field, i. e., $\boldsymbol{n} \sim \nabla \operatorname{dist}(., \partial\Omega)$. Due to the boundary condition $\boldsymbol{n} \times \boldsymbol{E} = \boldsymbol{0}$ we have $\beta(\boldsymbol{y}) = 0$ for all $\boldsymbol{y} \in \partial\Omega$. For $\boldsymbol{w}$ we then obtain

$$\boldsymbol{n}(\boldsymbol{x}_0) \cdot \boldsymbol{w}(\boldsymbol{x}_0) = \boldsymbol{n}(\boldsymbol{x}_0) \cdot \Big(\alpha(\boldsymbol{x}_0)\nabla \times \boldsymbol{n}(\boldsymbol{x}_0) + \beta(\boldsymbol{x}_0)\nabla \times \boldsymbol{t}(\boldsymbol{x}_0)$$

$$+ \nabla\alpha(\boldsymbol{x}_0) \times \boldsymbol{n}(\boldsymbol{x}_0) + \nabla\beta(\boldsymbol{x}_0) \times \boldsymbol{t}(\boldsymbol{x}_0)\Big)$$

$$= \nabla\alpha(\boldsymbol{x}_0) \cdot (\boldsymbol{n}(\boldsymbol{x}_0) \times \boldsymbol{n}(\boldsymbol{x}_0)) + \boldsymbol{t}(\boldsymbol{x}_0) \cdot (\boldsymbol{n}(\boldsymbol{x}_0) \times \nabla\beta(\boldsymbol{x}_0)) = 0.$$

The last term vanishes since $\nabla\beta(\boldsymbol{x}_0)$ is normal. $\qquad\square$

Theorem 2.1.25 (Regularity for time-harmonic fields). *Let $\boldsymbol{E} \in X_{\mathrm{N}}(\Omega; \varepsilon_r)$ and $\boldsymbol{H} \in X_{\mathrm{T}}(\Omega; 1)$ be solutions of the time-harmonic Maxwell equations (1.12)–(1.15). Then $\boldsymbol{E} \in H^{(r',r'')}(\mathrm{curl}, \Omega)$ with $r' > 0$, $r'' > 1/2$ and $\boldsymbol{H} \in H^{s'}(\Omega)^3$ with $s' > 1/2$, $\nabla \times \boldsymbol{H} \in H^{s''}(K)^3$ with $s'' > 0$ for all $K \subset \Omega$ where ε_r is constant.*

Proof. $\boldsymbol{E}$ is treated as in Theorem 2.1.24. For $\boldsymbol{H}$ we proceed as in the proof of Theorem 2.1.24. Since $\boldsymbol{H} \in X_{\mathrm{T}}(\Omega; 1)$, we can apply Theorem 1.2.1 to get $\boldsymbol{H} \in H^{s'}(\Omega)^3$ for some $s' > 1/2$. Now define $\boldsymbol{w} := 1/\varepsilon_r \nabla \times \boldsymbol{H}$. Clearly, $\nabla \times \boldsymbol{w} \in L^2(\Omega)^3$, $\nabla \cdot (\varepsilon_r\boldsymbol{w}) = 0$ in Ω, and $\boldsymbol{n} \times \boldsymbol{w} = \boldsymbol{0}$ on $\partial\Omega$ from (1.12). Hence $\boldsymbol{w} \in H^{s''}(\Omega)^3$ for some $s'' > 0$. This implies $\nabla \times \boldsymbol{H} \in H^{s''}(K)^3$ for all subsets K of Ω where ε_r is constant. $\qquad\square$

Remark 2.1.26.

(1) The assertion of the previous theorem applies to the nontrivial eigenfunctions of the eigenvalue problem, or to the solutions to the time-harmonic source problem with a suitable current $\boldsymbol{J}$.

(2) The exponent r' in Theorem 2.1.24 can be arbitrarily small, depending on $\partial\Omega$ and the jump discontinuity of ε_r [14]. If ε_r is Lipschitz-continuous (which

includes the case when ε_r is constant), then $r' > 1/2$ and if moreover Ω is convex, then $r' = 1$, as already denoted in Theorem 2.1.2. If ε_r is piecewise constant on two connected subdomains, then the electric field is piecewise in $(H^{r'})^3$ for $r' \geq 1/4$ and, if Ω is convex, for $r' \geq 1/2$ [14, Sect. 7a]. This holds especially in our applications for photonic crystals, Section 2.2.

The Nédélec edge elements

We now present a family of finite element methods that fulfills the requirements. Take $k \in \mathbb{N}$ and let $\mathcal{P}_k$ be the space of polynomials up to degree k, $\mathcal{P}'_k$ those of homogeneous degree k (i. e. $\mathcal{P}_k = \cup_{l=0}^{k} \mathcal{P}'_l$) and $\mathcal{P}_k^3$ the vector fields with components in $\mathcal{P}_k$. Now one defines

$$\widehat{\boldsymbol{V}}_k := \mathcal{P}_{k-1}^3 \oplus \{\boldsymbol{v} \in (\mathcal{P}'_k)^3 \, : \, \boldsymbol{x} \cdot \boldsymbol{v}(\boldsymbol{x}) = 0 \text{ for all } \boldsymbol{x} \in \mathbb{R}^3\}.$$

This polynomial space satisfies

$$\mathcal{P}_k^3 = \widehat{\boldsymbol{V}}_k \oplus \boldsymbol{\nabla}(\mathcal{P}'_{k+1})^3$$

and for $\boldsymbol{v} \in \widehat{\boldsymbol{V}}_k$ with $\boldsymbol{\nabla} \times \boldsymbol{v} = \boldsymbol{0}$ one gets $\boldsymbol{v} = \boldsymbol{\nabla} q$ for some $q \in \mathcal{P}_{k+1}$.

We let Σ_3 be the standard simplex in $\mathbb{R}^3$ and we consider R_k to be the finite element space on Σ^3. Let $\mathcal{T}$ be the decomposition of Ω into tetrahedra and define for $T \in \mathcal{T}$ a affine linear bijective mapping $\boldsymbol{F}_T : \Sigma_3 \to T$ with $\det(F_T) > 0$. The local finite element space $\boldsymbol{V}_T^h$ on T is then defined by

$$\boldsymbol{V}_T^h := \{\boldsymbol{v} = (D\boldsymbol{F}_T^{\dagger})^{-1}\widehat{\boldsymbol{v}} \circ \boldsymbol{F}_T^{-1} \, : \, \widehat{\boldsymbol{v}} \in \widehat{\boldsymbol{V}}_k\}.$$

A consequence of this definition is that $\boldsymbol{\nabla} \times \boldsymbol{v} = \boldsymbol{0}$ on T is equivalent to $\widehat{\boldsymbol{\nabla}} \times \widehat{\boldsymbol{v}} = \boldsymbol{0}$ on Σ_3. Our global finite element space is now given as

$$\boldsymbol{V}^h := \{\boldsymbol{v} \in H(\text{curl}, \Omega) \, : \, \boldsymbol{v}|_T \in \boldsymbol{V}_T^h\}.$$

This will give $\boldsymbol{V}^h \subset H(\text{curl}, \Omega)$ and that the boundary condition $\boldsymbol{n} \times \boldsymbol{v}^h = \boldsymbol{0}$ can be fulfilled. A more detailed description can be found in [26, Ch. 5.5]. The construction of corresponding elements on quadrilaterals is given in [26, Ch. 6.3]. The lowest order method is also described and used in Chapter 2.2.3.

There is a second family of curl-conforming elements on tetrahedra where the local space is $\mathcal{P}_k^3$ [26, Ch. 8.2]. Although they have more degrees of freedom, they do not improve the convergence order in the curl-norm for a given degree. However, they show the improved L^2-convergence which the first family does not provide.

Theorem 2.1.27 (Strong approximation of Nédélec elements 1). *There exists an interpolation operator* $\boldsymbol{\Pi}^h \, : \, H^{s_0}(\Omega)^3 \to \boldsymbol{V}^h$ *that satisfies the conditions of Assumption 2.1.20 for $s_0 > 1/2$.*

Proof. It is shown in [26, Thm. 5.41] that for $r = 1/2 + \delta$, $\delta > 0$, one gets

$$\|v - \boldsymbol{\Pi}^h v\|_{H(\mathrm{curl},T)} \leq C h_T^{1/2} \big(h_T^\delta \|v\|_{H^{1/2+\delta}(\omega_T)} + h_T^{1/2+\delta} \|\boldsymbol{\nabla} \times v\|_{H^\delta(\omega_T)} \big)$$
$$\leq C h_T^r \big(\|v\|_{H^r(\omega_T)} + h_T^{1/2} \|\boldsymbol{\nabla} \times v\|_{H^{r-1/2}(\omega_T)} \big).$$

Finally this yields for $r > 1/2$

$$\|v - \boldsymbol{\Pi}^h v\|_V \leq C h_T^r \|v\|_{H^{(r,r-1/2)}(\mathrm{curl},\Omega)}.$$

The remaining properties follow from [26, Ch. 5.5]. $\square$

Theorem 2.1.28 (Strong approximation of Nédélec elements 2). *There exists an interpolation operator* $\boldsymbol{\Pi}^h : L^2(\Omega)^3 \to V^h$ *that satisfies the conditions of Assumption 2.1.20 for some* $s_0 > 0$.

Proof. It has been shown in [27] that there exists an interpolation operator $\boldsymbol{\Pi}^h$: $L^2(\Omega)^3 \to V^h$ with $\|\boldsymbol{\Pi}^h v\|_{L^2(\Omega)^3} \leq C \|v\|_{L^2(\Omega)^3}$, $\|v - \boldsymbol{\Pi}^h v\|_{L^2(\Omega)^3} \leq C \|v\|_{H(\mathrm{curl},\Omega)}$ and $\|v - \boldsymbol{\Pi}^h v\|_{L^2(\Omega)^3} \leq C h \|v\|_{H^1(\mathrm{curl},\Omega)}$. By space interpolation [8, Ch. 14.2] we thus get $\|v - \boldsymbol{\Pi}^h v\|_{L^2(\Omega)^3} \leq C h^s \|v\|_{H^s(\mathrm{curl},\Omega)}$ for $s \in (0,1)$ and $h := \max\{h_T : T \in \mathcal{T}\}$. $\square$

Theorem 2.1.29 (Strong approximation of Nédélec elements 3). *Assume that* Ω *is decomposed into two disjoint connected subsets* Ω_1, Ω_2 *and* $\mathcal{T}$ *is a decomposition of* Ω *with either* $T \subset \Omega_1$ *or* $T \subset \Omega_2$ *for all* $T \in \mathcal{T}$. *Assume that* v *is in* $H^{(r,r-1/2)}(\Omega_1)^3 \cap H^{(r,r-1/2)}(\Omega_2)^3$. *Then Assumption 2.1.20 holds true with a set* ω_T *that is either in* Ω_1 *or* Ω_2 *in both cases* [6, p. 1286].

Proof. Both interpolation operators in Theorems 2.1.27 and 2.1.28 allow local estimates where there is the freedom to choose ω_T as required in the given situation. $\square$

Theorem 2.1.30 (Eigenvalue approximation in a periodic domain). *Assume that* ε_{r} *is piecewise constant and two-valued on a decomposition of* $\Omega := [0,1]^3$ *and* $\mathcal{T}$ *be a discretisation, both as in Theorem 2.1.29. If we discretise the electric or magnetic eigenvalue problem* (1.17), (1.16) *by Nédélec edge elements, the discrete eigenvalues will converge to the exact eigenvalues for decreasing meshsize. The convergence rate for the eigenvalues approximation will be at least 1.*

Proof. Electric field: Since Ω is convex and $\partial\Omega = \emptyset$ we have by Remark 2.1.26 that $\boldsymbol{E}$ is piecewise in $(H^{r'})^3$ for $r' \geq 1/2$ and by Theorem 2.1.24 that $\boldsymbol{\nabla} \times \boldsymbol{E} \in H^{r''}(\Omega)^3$ for $r'' > 1/2$. The result follows since Theorems 2.1.28 and 2.1.29 pave the way to apply Theorem 2.1.16. Note that the eigenvalue convergence is twice as large as the rate for the spatial approximation.

Magnetic field: We find solutions $\boldsymbol{H} \in H^{s'}(\Omega)^3$ and $\boldsymbol{\nabla} \times \boldsymbol{H} \in H^{s''}(\Omega_1)^3 \cap H^{s''}(\Omega_2)^3$ for $s' > 1/2$, $s'' > 0$. The results follow in the same way as above from Theorem 2.1.29. $\square$

2.1.6 Some Generalisations and Extensions

Collective compactness / Discrete compactness property

Convergence of the eigenvalues as in Theorem 2.1.12 (but without a rate) can also be obtained under the weaker condition that the family of discrete operators over a family of arbitrarily fine meshes is *collectively compact* [26, Ch. 2.3.3]. This can be reformulated in terms of a *discrete compactness property* of a family of approximation spaces [26, Ch. 7.3.2] and has been developed and used by many authors to prove convergence of different methods and elements, e. g. [22] [1] [10] [15] [5] [2, Sect. 19] and others.

H^1–conforming finite elements

Standard H^1-conforming spaces cannot be used in general, as already mentioned in Section 2.1.4, since solutions of Maxwell's equation can exhibit strong singularities. However, it was found out that formulating the divergence constraint in a weaker space than $L^2(\Omega)^3$ can cure the problem. Let Y be a Sobolev space in between $H^{-1}(\Omega)^3$ and $L^2(\Omega)^3$, $\boldsymbol{V}_h$ and Q_h be H^1-conforming finite element spaces of equal polynomial degree.

In the *Weighted Residual Method* [13] one considers $\boldsymbol{u}$ such that $\boldsymbol{\nabla}\cdot\boldsymbol{u} \in Y :=$ $\{\phi \in L^2(\Omega) : \|d^\gamma\phi\|_{L^2(\Omega)} < \infty\}$, where d a distance function from the singularities and γ some suitable positive number. One variant of the method is to seek $\boldsymbol{u}^h$ such that

$$\int_\Omega \{\boldsymbol{\nabla}\times\boldsymbol{u}^h \cdot \boldsymbol{\nabla}\times\boldsymbol{v}^h + d^{2\gamma}\boldsymbol{\nabla}\cdot\boldsymbol{u}^h\boldsymbol{\nabla}\cdot\boldsymbol{v}^h + \nabla p_h \cdot \boldsymbol{v}^h\} = \int_\Omega \boldsymbol{f}\cdot\boldsymbol{v}^h \quad \text{for all } \boldsymbol{v}^h \in \boldsymbol{V}_h$$

$$\int_\Omega \boldsymbol{u}^h \cdot \nabla q_h = 0 \qquad \text{for all } q_h \in Q_h.$$

The $H^{-\alpha}$-*regularisation* considers $\boldsymbol{\nabla}\cdot\boldsymbol{u} \in Y := H^{-\alpha}(\Omega)$ for some $\alpha \in [1/2, 1]$ and in order to get computable expressions one uses grid-dependent norms [7]. Here we seek $\boldsymbol{u}^h$ such that

$$\int_\Omega \{\boldsymbol{\nabla}\times\boldsymbol{u}^h \cdot \boldsymbol{\nabla}\times\boldsymbol{v}^h + h^{2\alpha}\boldsymbol{\nabla}\cdot\boldsymbol{u}^h\boldsymbol{\nabla}\cdot\boldsymbol{v}^h + \nabla p_h \cdot \boldsymbol{v}^h\} = \int_\Omega \boldsymbol{f}\cdot\boldsymbol{v}^h \quad \text{for all } \boldsymbol{v}^h \in \boldsymbol{V}_h$$

$$\int_\Omega \{-\boldsymbol{u}^h \cdot \nabla q_h + h^{2(1-\alpha)}\nabla q_h \cdot \nabla q_h\} = 0 \qquad \text{for all } q_h \in Q_h.$$

An overview on this topic is given in [11]. Note that this theory follows also the scheme presented in Assumptions 2.1.13.

Differential forms

In the assumptions of this section we used a structural condition concerning $\boldsymbol{\nabla}Q$ and $\boldsymbol{V}$ (see Theorem 2.1.4 (proof), Theorem 2.1.8, or Assumption 2.1.20). This comes naturally if one considers the electric and the magnetic field as differential forms. The underlying structure ("exact sequence") should also show up in the discrete spaces ("commuting diagram"). More on this topic can be found in [19] [2, Part. IV].

2.2   Calculation of the photonic band structure

by *Christian Wieners*

2.2.1   The parameterised periodic eigenvalue problem

We consider the Maxwell eigenvalue problem for the magnetic field for optic waves in a periodic medium in $\mathbb{R}^3$ with $\mu_r \equiv 1$ for the magnetic permeability (so that there is no magnetic induction in the medium), and assume that the electric permittivity ε_r is a periodic function on the lattice $\Gamma := \mathbb{Z}^3$ (that is, $\varepsilon_r(\boldsymbol{x}) = \varepsilon_r(\boldsymbol{x} + \boldsymbol{R})$ for all $\boldsymbol{R} \in \mathbb{Z}^3$). This eigenvalue problem has the following form: determine eigenfunctions $\boldsymbol{H} \colon \mathbb{R}^3 \longrightarrow \mathbb{C}^3$ and eigenvalues $\lambda \in \mathbb{C}$ such that

$$\nabla \times \left(\varepsilon_r^{-1} \nabla \times \boldsymbol{H} \right) \;=\; \lambda \boldsymbol{H}, \qquad\qquad (2.17\text{a})$$
$$\nabla \cdot \boldsymbol{H} \;=\; 0, \qquad\qquad (2.17\text{b})$$

in $\mathbb{R}^3$ (cf. Chapter 1.1.5).

Therefore, we proceed as in Chapter 1.1.7. If we take $\Omega := (0,1)^3$ to be the fundamental cell of the periodic structure, we have to find for any $\boldsymbol{k} \in \mathbb{R}^3$ a series of (quasi-periodic) eigenfunctions $\boldsymbol{H}$ that can be represented as

$$\boldsymbol{H}(\boldsymbol{x}) \;=\; \mathrm{e}^{\mathrm{i}\boldsymbol{k}\cdot\boldsymbol{x}} \widetilde{\boldsymbol{H}}(\boldsymbol{x}) \qquad \text{for all } \boldsymbol{x} \in \Omega, \qquad\qquad (2.18)$$

where $\widetilde{\boldsymbol{H}}$ is now a periodic function in the fundamental cell Ω. The curl-operator applied to this ansatz yields

$$\nabla \times \boldsymbol{H}(\boldsymbol{x}) = \mathrm{e}^{\mathrm{i}\boldsymbol{k}\cdot\boldsymbol{x}} \left(\nabla \times \widetilde{\boldsymbol{H}}(\boldsymbol{x}) + \mathrm{i}\boldsymbol{k} \times \widetilde{\boldsymbol{H}}(\boldsymbol{x}) \right) = \mathrm{e}^{\mathrm{i}\boldsymbol{k}\cdot\boldsymbol{x}} \nabla_{\boldsymbol{k}} \times \widetilde{\boldsymbol{H}}(\boldsymbol{x}).$$

Here, we have introduced the *shifted gradient operator*

$$\nabla_{\boldsymbol{k}} \;=\; \nabla + \mathrm{i}\boldsymbol{k} \qquad\qquad (2.19)$$

and we consider the transformed eigenvalue problem: determine periodic eigenfunctions $\widetilde{\boldsymbol{H}} \colon \Omega \longrightarrow \mathbb{C}^3$ and eigenvalues $\lambda \in \mathbb{C}$ satisfying

$$\nabla_{\boldsymbol{k}} \times \left(\varepsilon_r^{-1} \nabla_{\boldsymbol{k}} \times \widetilde{\boldsymbol{H}} \right) \;=\; \lambda \widetilde{\boldsymbol{H}}, \qquad\qquad (2.20\text{a})$$
$$\nabla_{\boldsymbol{k}} \cdot \widetilde{\boldsymbol{H}} \;=\; 0 \qquad\qquad (2.20\text{b})$$

in Ω. Then, defining $\boldsymbol{H}$ by (2.18) indeed gives an eigenfunction of (2.17).

We recall (Theorem 2.1.7) that the spectrum is real and discrete with ordered real eigenvalues

$$0 \leq \lambda_{\boldsymbol{k}}^1 \leq \lambda_{\boldsymbol{k}}^2 \leq \cdots \leq \lambda_{\boldsymbol{k}}^N \leq \cdots .$$

The Floquet–Bloch theory guarantees that the full spectrum of the Maxwell operator associated to problem (2.17) can be computed by a series of eigenvalue problems (2.20) in the fundamental cell $\Omega = (0,1)^3$ of the periodic structure, where it is sufficient that $\boldsymbol{k}$ is in the Brillouin zone $B = [-\pi, \pi]^3$. The spectrum of the problem (2.17) is given by

$$\sigma = \bigcup_{n \in \mathbb{N}} [\inf_{\boldsymbol{k} \in B} \lambda_{\boldsymbol{k}}^n, \sup_{\boldsymbol{k} \in B} \lambda_{\boldsymbol{k}}^n]$$

and possible nonempty intervals

$$(\sup_{\boldsymbol{k} \in B} \lambda_{\boldsymbol{k}}^n, \inf_{\boldsymbol{k} \in B} \lambda_{\boldsymbol{k}}^{n+1})$$

are called band gaps. In the following we develop a numerical method to compute photonic bandstructures to find such bandgaps.

2.2.2 Galerkin approximation of the eigenvalue problems

In the next step, we derive a weak formulation for the eigenvalue problem (2.20). This allows for general coefficients $\varepsilon_{\mathrm{r}} \in L^\infty(\Omega)$ with $1 \leq \varepsilon_{\mathrm{r}}(\boldsymbol{x}) \leq \varepsilon_\infty$ for almost all $\boldsymbol{x} \in \Omega$. In principle, it is also possible to consider a tensor-valued electric permittivity ε_{r} (as it is required, e. g., for anisotropic materials), but in our examples we will consider only the special case that we have a material with $\varepsilon_\infty > 1$ in a subset $\omega \subset \Omega$ and empty space in $\Omega \setminus \omega$, where we set $\varepsilon_{\mathrm{r}} = 1$.

We use the spaces

$$\boldsymbol{V} = H_{\mathrm{per}}(\mathrm{curl}, \Omega) := \left\{ \boldsymbol{u}|_\Omega : \boldsymbol{u} \in H(\mathrm{curl}, \mathbb{R}^3/\mathbb{Z}^3) \right\},$$

$$Q = H_{\mathrm{per}}^1(\Omega) := \left\{ q|_\Omega : q \in H^1(\mathbb{R}^3/\mathbb{Z}^3) \right\}$$

of periodic functions on $\Omega = (0,1)^3$ (that is, with periodic boundary conditions on $\partial\Omega$ for the corresponding trace operator), see Chapter 1.2.6. For $\boldsymbol{u}, \boldsymbol{v} \in \boldsymbol{V}$ and $p, q \in Q$ we define the Hermitian sesqui-linear forms[1]

$$a_{\boldsymbol{k}}(\boldsymbol{u}, \boldsymbol{v}) = \int_\Omega \varepsilon_{\mathrm{r}}^{-1} \overline{\boldsymbol{\nabla}_{\boldsymbol{k}} \times \boldsymbol{u}} \cdot \boldsymbol{\nabla}_{\boldsymbol{k}} \times \boldsymbol{v} \, d\boldsymbol{x} \,,$$

$$m(\boldsymbol{u}, \boldsymbol{v}) = \int_\Omega \overline{\boldsymbol{u}} \cdot \boldsymbol{v} \, d\boldsymbol{x} \,,$$

---

[1] Note that in our notation the sesqui-linear forms (and also all dual pairings) are anti-linear in the first component.

$$b_{\boldsymbol{k}}(\boldsymbol{v}, q) = \int_{\Omega} \overline{\boldsymbol{v}} \cdot \boldsymbol{\nabla}_{\boldsymbol{k}} q \, dx \,,$$

$$c_{\boldsymbol{k}}(p, q) = \int_{\Omega} \overline{\boldsymbol{\nabla}_{\boldsymbol{k}} p} \cdot \boldsymbol{\nabla}_{\boldsymbol{k}} q \, dx \,,$$

and we define — depending on the Floquet parameter $\boldsymbol{k}$ — the subspace which includes the $\boldsymbol{k}$-divergence constraint (2.20b) by

$$\boldsymbol{X}_{\boldsymbol{k}} \;\; = \big\{ \boldsymbol{v} \in \boldsymbol{V} : b_{\boldsymbol{k}}(\boldsymbol{v}, q) = 0 \text{ for all } q \in Q \big\}$$

($\boldsymbol{X}_{\boldsymbol{0}} = \boldsymbol{V}_1$ in (2.7)). Now, we can rewrite the equation (2.20) in weak form: find $(\boldsymbol{u}, \lambda) \in \boldsymbol{X}_{\boldsymbol{k}} \setminus \{0\} \times \mathbb{R}$ such that

$$a_{\boldsymbol{k}}(\boldsymbol{u}, \boldsymbol{v}) = \lambda \, m(\boldsymbol{u}, \boldsymbol{v}) \qquad \text{for all } \boldsymbol{v} \in \boldsymbol{X}_{\boldsymbol{k}} \,. \tag{2.21}$$

Obviously, $\boldsymbol{\nabla}_{\boldsymbol{k}} Q$ is contained in the kernel of the sesqui-linear form $a_{\boldsymbol{k}}(\cdot, \cdot)$, i.e., $a_{\boldsymbol{k}}(\boldsymbol{v}, \boldsymbol{v}) = 0$ for $\boldsymbol{v} = \boldsymbol{\nabla}_{\boldsymbol{k}} q$ with $q \in Q$. Moreover, for any $\delta > 0$, the sesqui-linear form

$$a_{\boldsymbol{k}}^{\delta}(\boldsymbol{u}, \boldsymbol{v}) = a_{\boldsymbol{k}}(\boldsymbol{u}, \boldsymbol{v}) + \delta \, m(\boldsymbol{u}, \boldsymbol{v}) \tag{2.22}$$

is elliptic in $\boldsymbol{V}$ (and thus also in $\boldsymbol{X}_{\boldsymbol{k}}$). Since $\boldsymbol{X}_{\boldsymbol{k}}$ embeds compactly into $L^2_{\text{per}}(\Omega)^3$ (see Theorem 1.2.1 and Theorem 2.1.7), the spectrum of the corresponding self-adjoint operator is discrete, positive, and all eigenvalues have finite multiplicity. As a consequence, the eigenvalue problem (2.21) has a discrete non-negative spectrum, and the dimension of kernel of the sesqui-linear form $a_{\boldsymbol{k}}(\cdot, \cdot)$ in $\boldsymbol{X}_{\boldsymbol{k}}$ is finite.

For the discrete Galerkin approximation of (2.21) we consider finite element subspaces $\boldsymbol{V}_{h,\boldsymbol{k}} \subset \boldsymbol{V}$ and $Q_{h,\boldsymbol{k}} \subset Q$, and we define the constrained discrete space

$$\boldsymbol{X}_{h,\boldsymbol{k}} \;\; = \;\; \big\{ \boldsymbol{v}_h \in \boldsymbol{V}_h : b_{\boldsymbol{k}}(\boldsymbol{v}_h, q_h) = 0 \text{ for all } q_h \in Q_h \big\} \,.$$

Note that in $\boldsymbol{X}_{h,\boldsymbol{k}}$ the $\boldsymbol{k}$-divergence constraint is fulfilled only approximately, so that the discretisation is non-conforming, i.e., we have $\boldsymbol{X}_{h,\boldsymbol{k}} \not\subset \boldsymbol{X}_{\boldsymbol{k}}$.

On the discrete space, we define the Galerkin approximation: find $(\boldsymbol{u}_{h,\boldsymbol{k}}, \lambda_{h,\boldsymbol{k}}) \in \boldsymbol{X}_{h,\boldsymbol{k}} \setminus \{0\} \times \mathbb{R}$ such that

$$a_{\boldsymbol{k}}(\boldsymbol{u}_{h,\boldsymbol{k}}, \boldsymbol{v}_{h,\boldsymbol{k}}) = \lambda_{h,\boldsymbol{k}} \, m(\boldsymbol{u}_{h,\boldsymbol{k}}, \boldsymbol{v}_{h,\boldsymbol{k}}) \qquad \text{for all } \boldsymbol{v}_{h,\boldsymbol{k}} \in \boldsymbol{X}_{h,\boldsymbol{k}} \,. \tag{2.23}$$

Due to the nonconformity of the discretisation, it is not a priori clear that all eigenvalues of the discrete problem indeed approximate the eigenvalues of the continuous problem, compare Chapter 2.1.4. This requires a suitable choice of the finite element spaces.

For the numerical analysis of the eigenvalue problem (2.23) we refer to [4]. If the Assumptions 2.1.13 and 2.1.20 can be verified (as in the previous section for $\boldsymbol{k} = \boldsymbol{0}$), we can apply Theorem 2.1.16 which guarantees eigenvalue convergence.

2.2.3 Lowest order conforming finite elements on hexahedral meshes

We now introduce a family of finite element spaces which satisfy the required conditions for eigenvalue convergence [4].

Let $\bar{\Omega} = \cup_{c \in \mathcal{C}_h} \bar{\Omega}_c$ be a decomposition into hexahedral cells $c \in \mathcal{C}_h$, and let $\varphi_c \colon \hat{\Omega} = [0,1]^3 \longrightarrow \Omega_c$ be the transformation to the reference cell. For $\boldsymbol{k} = \boldsymbol{0}$ we use the standard Nédélec elements with edge degrees of freedom [26, Chap. 6.3] and standard trilinear conforming H^1 elements, i. e.,

$$\boldsymbol{V}_{h,\boldsymbol{0}} = \{\boldsymbol{u} \in \boldsymbol{X_0} \colon D\varphi_c^T \boldsymbol{u} \circ \varphi_c \in \mathcal{P}_{0,1,1}\boldsymbol{e}_x + \mathcal{P}_{1,0,1}\boldsymbol{e}_y + \mathcal{P}_{1,1,0}\boldsymbol{e}_z \text{ for all } c \in \mathcal{C}_h\},$$
$$Q_{h,\boldsymbol{0}} = \{q \in Q \colon q \circ \varphi_c \in \mathcal{P}_{1,1,1} \text{ for all } c \in \mathcal{C}_h\},$$

where $\mathcal{P}_{i,j,k}$ denotes the space of polynomials with degree i, j, k in coordinate x_1, x_2, x_3, respectively. For the implementation we use a nodal basis in both cases, where each basis function is associated to a nodal point.

Let $\{\boldsymbol{z}_v \colon v \in \mathcal{V}_h\} \subset \bar{\Omega}$ be the set of all vertices, where $\mathcal{V}_h$ denotes an index set enumerating the vertices, and let $\mathcal{E}_h$ be the set of all edges $e = (\boldsymbol{x}_e, \boldsymbol{y}_e)$, i. e., e is understood as an ordered tupel of two vertices. Note that this defines an orientation for every edge which is independent of the cells. The midpoint of e is defined by $\boldsymbol{z}_e = (\boldsymbol{x}_e + \boldsymbol{y}_e)/2$ and and its tangent by $\boldsymbol{t}_e = \boldsymbol{y}_e - \boldsymbol{x}_e$.

In $\boldsymbol{V}_{h,\boldsymbol{0}}$ a nodal basis function $\boldsymbol{\psi}_{e,\boldsymbol{0}}$ is associated to every edge $e \in \mathcal{E}_h$ such that for all $\boldsymbol{u}_{h,\boldsymbol{0}} \in \boldsymbol{V}_{h,\boldsymbol{0}}$

$$\boldsymbol{u}_{h,\boldsymbol{0}} = \sum_{e \in \mathcal{E}_h} \langle \boldsymbol{\psi}'_{e,\boldsymbol{0}}, \boldsymbol{u}_{h,\boldsymbol{0}} \rangle \boldsymbol{\psi}_{e,\boldsymbol{0}}, \qquad \langle \boldsymbol{\psi}'_{e,\boldsymbol{0}}, \boldsymbol{u}_{h,\boldsymbol{0}} \rangle = \int_{\boldsymbol{x}_e}^{\boldsymbol{y}_e} \boldsymbol{u}_{h,\boldsymbol{0}} \cdot \boldsymbol{t}_e \, ds \,.$$

Here, the dual functionals $\boldsymbol{\psi}'_{e,\boldsymbol{0}} \in \boldsymbol{V}'_{h,\boldsymbol{0}}$ denote the degree of freedom associated to the edge e. Note that due to the periodic boundary conditions the boundary degrees of freedom are not independent, i. e., we have $\boldsymbol{\psi}'_{e_1,\boldsymbol{0}} = \boldsymbol{\psi}'_{e_2,\boldsymbol{0}}$ if $\boldsymbol{z}_{e_1}, \boldsymbol{z}_{e_2} \in \partial\Omega$ with $\boldsymbol{z}_{e_1} - \boldsymbol{z}_{e_2} \in \mathbb{Z}^3$.

Analogously, we have nodal basis functions $\phi_{v,\boldsymbol{0}} \in Q_{h,\boldsymbol{0}}$ associated to a vertex v such that for all $q_{h,\boldsymbol{0}} \in Q_{h,\boldsymbol{0}}$

$$q_{h,\boldsymbol{0}} = \sum_{v \in \mathcal{V}_h} \langle \phi'_{v,\boldsymbol{0}}, q_{h,\boldsymbol{0}} \rangle \phi_{v,\boldsymbol{0}}, \qquad \langle \phi'_{v,\boldsymbol{0}}, q_{h,\boldsymbol{0}} \rangle = q_{h,\boldsymbol{0}}(\boldsymbol{z}_v) \,.$$

Again, the degrees of freedom $\phi'_{v,\boldsymbol{0}} \in Q_{h,\boldsymbol{0}}$ are periodic, i. e., $\phi'_{v_1,\boldsymbol{0}} = \phi'_{v_2,\boldsymbol{0}}$ and thus $q_{h,\boldsymbol{0}}(\boldsymbol{z}_{v_1}) = q_{h,\boldsymbol{0}}(\boldsymbol{z}_{v_2})$ for $\boldsymbol{z}_{v_1}, \boldsymbol{z}_{v_2} \in \partial\Omega$ with $\boldsymbol{z}_{v_1} - \boldsymbol{z}_{v_2} \in \mathbb{Z}^3$.

An important feature of the pair of finite element spaces $\boldsymbol{V}_{h,\boldsymbol{0}} \times Q_{h,\boldsymbol{0}}$ is the property $\nabla Q_{h,\boldsymbol{0}} \subset \boldsymbol{V}_{h,\boldsymbol{0}}$ (see Theorem 2.1.8), i. e., we have

$$\nabla q_{h,\boldsymbol{0}} = \sum_{e \in \mathcal{E}_h} \langle \boldsymbol{\psi}'_{e,\boldsymbol{0}}, \nabla q_{h,\boldsymbol{0}} \rangle \boldsymbol{\psi}_{e,\boldsymbol{0}} \tag{2.24}$$

for all $q_{h,\boldsymbol{0}} \in Q_{h,\boldsymbol{0}}$.

2.2.4   Conforming elements with phase-shifted nodal basis

Starting from the standard nodal basis and degrees of freedom, we now consider modified elements for $\boldsymbol{k} \neq \boldsymbol{0}$ with a *phase shift*: we define

$$
\begin{aligned}
\boldsymbol{V}_{h,\boldsymbol{k}} &= \operatorname{Span}\{\boldsymbol{\psi}_{e,\boldsymbol{k}} : e \in \mathcal{E}_h\}, & \boldsymbol{\psi}_{e,\boldsymbol{k}}(\boldsymbol{x}) &= \mathrm{e}^{-\mathrm{i}\boldsymbol{k}\cdot(\boldsymbol{x}-\boldsymbol{z}_e)}\boldsymbol{\psi}_{e,\boldsymbol{0}}(\boldsymbol{x}), \\
Q_{h,\boldsymbol{k}} &= \operatorname{Span}\{\phi_{v,\boldsymbol{k}} : v \in \mathcal{V}_h\}, & \phi_{v,\boldsymbol{k}}(\boldsymbol{x}) &= \mathrm{e}^{-\mathrm{i}\boldsymbol{k}\cdot(\boldsymbol{x}-\boldsymbol{z}_v)}\phi_{v,\boldsymbol{0}}(\boldsymbol{x})
\end{aligned}
$$

with the basis representation depending on the phase-shifted degrees of freedom

$$
\begin{aligned}
\boldsymbol{u}_{h,\boldsymbol{k}} &= \sum_{e\in\mathcal{E}_h} \langle \boldsymbol{\psi}'_{e,\boldsymbol{k}}, \boldsymbol{u}_{h,\boldsymbol{k}}\rangle \boldsymbol{\psi}_{e,\boldsymbol{k}}, & \langle \boldsymbol{\psi}'_{e,\boldsymbol{k}}, \boldsymbol{u}_{h,\boldsymbol{k}}\rangle &= \int_{\boldsymbol{x}_e}^{\boldsymbol{y}_e} \mathrm{e}^{\mathrm{i}\boldsymbol{k}\cdot(\boldsymbol{x}-\boldsymbol{z}_e)}\boldsymbol{u}_{h,\boldsymbol{0}}\cdot\boldsymbol{t}_e\,ds, \\
q_{h,\boldsymbol{k}} &= \sum_{v\in\mathcal{V}_h} \langle \phi'_{v,\boldsymbol{k}}, q_{h,\boldsymbol{k}}\rangle \phi_{v,\boldsymbol{k}}, & \langle \phi'_{v,\boldsymbol{k}}, q_{h,\boldsymbol{0}}\rangle &= q_{h,\boldsymbol{k}}(\boldsymbol{z}_v).
\end{aligned}
$$

In the implementation we use the transformation properties of the phase-shifted basis functions: we have

$$
\begin{aligned}
\boldsymbol{\nabla}_{\boldsymbol{k}}\,\phi_{v,\boldsymbol{k}}(\boldsymbol{x}) &= \mathrm{e}^{-\mathrm{i}\boldsymbol{k}\cdot(\boldsymbol{x}-\boldsymbol{z}_v)}\boldsymbol{\nabla}\phi_{v,\boldsymbol{0}}(\boldsymbol{x}), \\
\boldsymbol{\nabla}_{\boldsymbol{k}}\times\boldsymbol{\psi}_{e,\boldsymbol{k}}(\boldsymbol{x}) &= \mathrm{e}^{-\mathrm{i}\boldsymbol{k}\cdot(\boldsymbol{x}-\boldsymbol{z}_e)}\boldsymbol{\nabla}\times\boldsymbol{\psi}_{e,\boldsymbol{0}}(\boldsymbol{x}), \\
\boldsymbol{\nabla}_{\boldsymbol{k}}\cdot\boldsymbol{\psi}_{e,\boldsymbol{k}}(\boldsymbol{x}) &= \mathrm{e}^{-\mathrm{i}\boldsymbol{k}\cdot(\boldsymbol{x}-\boldsymbol{z}_e)}\boldsymbol{\nabla}\cdot\boldsymbol{\psi}_{e,\boldsymbol{0}}(\boldsymbol{x}).
\end{aligned}
$$

As a consequence, all sesqui-linear forms are obtained by simple scaling, i. e.,

$$
\begin{aligned}
a_{\boldsymbol{k}}(\boldsymbol{\psi}_{e_1,\boldsymbol{k}}, \boldsymbol{\psi}_{e_2,\boldsymbol{k}}) &= \mathrm{e}^{-\mathrm{i}\boldsymbol{k}\cdot(\boldsymbol{z}_{e_1}-\boldsymbol{z}_{e_2})}\,a_{\boldsymbol{0}}(\boldsymbol{\psi}_{e_1,\boldsymbol{0}}, \boldsymbol{\psi}_{e_2,\boldsymbol{0}}), \\
m(\boldsymbol{\psi}_{e_1,\boldsymbol{k}}, \boldsymbol{\psi}_{e_2,\boldsymbol{k}}) &= \mathrm{e}^{-\mathrm{i}\boldsymbol{k}\cdot(\boldsymbol{z}_{e_1}-\boldsymbol{z}_{e_2})}\,m(\boldsymbol{\psi}_{e_1,\boldsymbol{0}}, \boldsymbol{\psi}_{e_2,\boldsymbol{0}}), \\
b_{\boldsymbol{k}}(\boldsymbol{\psi}_{e,\boldsymbol{k}}, \phi_{v,\boldsymbol{k}}) &= \mathrm{e}^{-\mathrm{i}\boldsymbol{k}\cdot(\boldsymbol{z}_e-\boldsymbol{z}_v)}\,b_{\boldsymbol{0}}(\boldsymbol{\psi}_{e,\boldsymbol{0}}, \phi_{v,\boldsymbol{0}}), \\
c_{\boldsymbol{k}}(\phi_{v_1,\boldsymbol{k}}, \phi_{v_2,\boldsymbol{k}}) &= \mathrm{e}^{-\mathrm{i}\boldsymbol{k}\cdot(\boldsymbol{z}_{v_1}-\boldsymbol{z}_{v_2})}\,c_{\boldsymbol{0}}(\phi_{v_1,\boldsymbol{0}}, \phi_{v_2,\boldsymbol{0}})
\end{aligned}
$$

for all $e, e_1, e_2 \in \mathcal{E}_h$ and $v, v_1, v_2 \in \mathcal{V}_h$. This property makes the use of the phase-shifted basis very attractive, and only small modifications from the standard elements are required in the implementation. In particular, inserting (2.24) we obtain

$$
\begin{aligned}
\boldsymbol{\nabla}_{\boldsymbol{k}}\,\phi_{v,\boldsymbol{k}}(\boldsymbol{x}) &= \mathrm{e}^{-\mathrm{i}\boldsymbol{k}\cdot(\boldsymbol{x}-\boldsymbol{z}_v)}\boldsymbol{\nabla}\phi_{v,\boldsymbol{0}}(\boldsymbol{x}) \\
&= \mathrm{e}^{-\mathrm{i}\boldsymbol{k}\cdot(\boldsymbol{x}-\boldsymbol{z}_v)}\sum_{e\in\mathcal{E}_h} \langle \boldsymbol{\psi}'_{e,\boldsymbol{0}}, \boldsymbol{\nabla}\phi_{v,\boldsymbol{0}}\rangle \boldsymbol{\psi}_{e,\boldsymbol{0}}(\boldsymbol{x}) \\
&= \mathrm{e}^{\mathrm{i}\boldsymbol{k}\cdot\boldsymbol{z}_v}\sum_{e\in\mathcal{E}_h} \langle \boldsymbol{\psi}'_{e,\boldsymbol{k}}, \boldsymbol{\nabla}\phi_{v,\boldsymbol{0}}\rangle \mathrm{e}^{-\mathrm{i}\boldsymbol{k}\cdot\boldsymbol{z}_e}\boldsymbol{\psi}_{e,\boldsymbol{k}}(\boldsymbol{x}),
\end{aligned}
$$

i. e., the inclusion $\boldsymbol{\nabla}_{\boldsymbol{k}}Q_{h,\boldsymbol{k}} \subset \boldsymbol{V}_{h,\boldsymbol{k}}$ is also satisfied for the phase-shifted elements.

The sesqui-linear form $c_{\boldsymbol{0}}(\cdot,\cdot)$ vanishes for all constants, i. e., $c_{\boldsymbol{0}}(\boldsymbol{v},\boldsymbol{v}) = 0$ if and only if $\boldsymbol{v}$ is constant. Thus, $c_{\boldsymbol{0}}(\cdot,\cdot)$ is elliptic in $\hat{Q}_{h,\boldsymbol{0}} = Q_{h,\boldsymbol{0}}/\mathbb{R}$. For $\boldsymbol{k} \neq \boldsymbol{0}$ we observe that the sesqui-linear form $c_{\boldsymbol{k}}(\cdot,\cdot)$ is elliptic in $Q_{h,\boldsymbol{k}}$, since the kernel of the operator $\boldsymbol{\nabla}_{\boldsymbol{k}}$ is spanned by non-periodic functions $\mathrm{e}^{\mathrm{i}\boldsymbol{k}\cdot\boldsymbol{x}}$. For $\boldsymbol{k} \neq \boldsymbol{0}$ we set $\hat{Q}_{h,\boldsymbol{k}} = Q_{h,\boldsymbol{k}}$ in order to unify the notation.

2.2.5 Discrete operators

For the formulation of the following algorithms we introduce discrete operators

$$
\begin{aligned}
A_{h,\mathbf{k}} &: & \mathbf{V}_{h,\mathbf{k}} &\longrightarrow \mathbf{V}'_{h,\mathbf{k}}, \\
M_{h,\mathbf{k}} &: & \mathbf{V}_{h,\mathbf{k}} &\longrightarrow \mathbf{V}'_{h,\mathbf{k}}, \\
B_{h,\mathbf{k}} &: & \mathbf{V}_{h,\mathbf{k}} &\longrightarrow Q'_{h,\mathbf{k}}, \\
C_{h,\mathbf{k}} &: & Q_{h,\mathbf{k}} &\longrightarrow Q'_{h,\mathbf{k}}
\end{aligned}
$$

induced by the sesqui-linear forms, i.e.,

$$
\begin{aligned}
\langle A_{h,\mathbf{k}}\boldsymbol{\psi}_{e_1,\mathbf{k}}, \boldsymbol{\psi}_{e_2,\mathbf{k}}\rangle &= a_{\mathbf{k}}(\boldsymbol{\psi}_{e_1,\mathbf{k}}, \boldsymbol{\psi}_{e_2,\mathbf{k}}), \\
\langle M_{h,\mathbf{k}}\boldsymbol{\psi}_{e_1,\mathbf{k}}, \boldsymbol{\psi}_{e_2,\mathbf{k}}\rangle &= m(\boldsymbol{\psi}_{e_1,\mathbf{k}}, \boldsymbol{\psi}_{e_2,\mathbf{k}}), \\
\langle B_{h,\mathbf{k}}\boldsymbol{\psi}_{e,\mathbf{k}}, \phi_{v,\mathbf{k}}\rangle &= b_{\mathbf{k}}(\boldsymbol{\psi}_{e,\mathbf{k}}, \phi_{v,\mathbf{k}}), \\
\langle C_{h,\mathbf{k}}\phi_{v_1,\mathbf{k}}, \phi_{v_2,\mathbf{k}}\rangle &= c_{\mathbf{k}}(\phi_{v_1,\mathbf{k}}, \phi_{v_2,\mathbf{k}})
\end{aligned}
$$

for all $e, e_1, e_2 \in \mathcal{E}_h$ and $v, v_1, v_2 \in \mathcal{V}_h$. In the implementation, the operators are represented by matrices $\underline{A}_{h,\mathbf{k}} = (\underline{A}_{e_1,e_2})_{e_1,e_2\in\mathcal{E}_h}$, $\underline{M}_{h,\mathbf{k}} = (\underline{M}_{e_1,e_2})_{e_1,e_2\in\mathcal{E}_h}$, $\underline{B}_{h,\mathbf{k}} = (\underline{B}_{v,e})_{v\in\mathcal{V}_h, e\in\mathcal{E}_h}$, $\underline{C}_{h,\mathbf{k}} = (\underline{C}_{v_1,v_2})_{v_1,v_2\in\mathcal{V}_h}$ with

$$
A_{h,\mathbf{k}} = \sum_{e_1,e_2\in\mathcal{E}_h} \underline{A}_{e_1,e_2} \boldsymbol{\psi}'_{e_2,\mathbf{k}} \otimes \boldsymbol{\psi}'_{e_1,\mathbf{k}}, \qquad \underline{A}_{e_1,e_2} = a_{\mathbf{k}}(\boldsymbol{\psi}_{e_2,\mathbf{k}}, \boldsymbol{\psi}_{e_1,\mathbf{k}}),
$$

$$
M_{h,\mathbf{k}} = \sum_{e_1,e_2\in\mathcal{E}_h} \underline{M}_{e_1,e_2} \boldsymbol{\psi}'_{e_2,\mathbf{k}} \otimes \boldsymbol{\psi}'_{e_1,\mathbf{k}}, \qquad \underline{M}_{e_1,e_2} = m(\boldsymbol{\psi}_{e_2,\mathbf{k}}, \boldsymbol{\psi}_{e_1,\mathbf{k}}),
$$

$$
B_{h,\mathbf{k}} = \sum_{v\in\mathcal{V}_h, e\in\mathcal{E}_h} \underline{B}_{v,e} \boldsymbol{\psi}'_{e,\mathbf{k}} \otimes \phi'_{v,\mathbf{k}}, \qquad \underline{B}_{v,e} = b_{\mathbf{k}}(\boldsymbol{\psi}_{e,\mathbf{k}}, \phi_{v,\mathbf{k}}),
$$

$$
C_{h,\mathbf{k}} = \sum_{v_1,v_2\in\mathcal{V}_h} \underline{C}_{v_1,v_2} \phi'_{v_2,\mathbf{k}} \otimes \phi'_{v_1,\mathbf{k}}, \qquad \underline{C}_{v_1,v_2} = c_{\mathbf{k}}(\phi_{v_2,\mathbf{k}}, \phi_{v_1,\mathbf{k}}),
$$

where the rank-one-operators are defined by $(\boldsymbol{\psi}'_1 \otimes \boldsymbol{\psi}'_2)\boldsymbol{u} = \langle \boldsymbol{\psi}'_1, \boldsymbol{u}\rangle \boldsymbol{\psi}'_2$. In all cases, the matrix representation is sparse, i.e., the number of non-zero entries is proportional to the number of edges and vertices. The matrix representations of $A_{h,\mathbf{k}}$ and $C_{h,\mathbf{k}}$ are used in the construction of smoothers for the multigrid iteration. A matrix representation of $B_{h,\mathbf{k}}$ can be avoided in the implementation since the operator can be directly evaluated by

$$
B_{h,\mathbf{k}}\boldsymbol{v}_{h,\mathbf{k}} = \sum_{v\in\mathcal{V}_h} b_{\mathbf{k}}(\boldsymbol{v}_{h,\mathbf{k}}, \phi_{v,\mathbf{k}})\phi'_{v,\mathbf{k}}
$$

and the coefficients can be computed by local integration. Nevertheless, using a matrix representation is far more efficient since this operation is applied very often.

The operator $B_{h,\mathbf{k}} \colon \mathbf{V}_{h,\mathbf{k}} \longrightarrow Q'_{h,\mathbf{k}}$ is the phase-shifted divergence operator in weak form, and its adjoint $B'_{h,\mathbf{k}} \colon Q_{h,\mathbf{k}} \longrightarrow \mathbf{V}'_{h,\mathbf{k}}$ corresponds to the weak phase-shifted gradient. It is an important observation for our algorithms, that the strong consistency property $\nabla_{\boldsymbol{k}} Q_{h,\boldsymbol{k}} \subset \boldsymbol{V}_{h,\boldsymbol{k}}$ of the discretisations allows also for a phase-shifted gradient operator in strong form

$$S_{h,\boldsymbol{k}} \colon Q_{h,\boldsymbol{k}} \longrightarrow \boldsymbol{V}_{h,\boldsymbol{k}}$$

which is defined by nodal evaluation, i. e.,

$$
\begin{aligned}
S_{h,\boldsymbol{k}} q_{h,\boldsymbol{k}} &= \sum_{e \in \mathcal{E}_h} \langle \boldsymbol{\psi}'_{e,\boldsymbol{k}}, \nabla_{\boldsymbol{k}} q_{h,\boldsymbol{k}} \rangle \boldsymbol{\psi}_{e,\boldsymbol{k}} \\
&= \sum_{e=(\boldsymbol{x}_e, \boldsymbol{y}_e) \in \mathcal{E}_h} \left(q_{h,\boldsymbol{k}}(\boldsymbol{y}_e) e^{i\boldsymbol{k} \cdot (\boldsymbol{y}_e - \boldsymbol{z}_e)} - q_{h,\boldsymbol{k}}(\boldsymbol{x}_e) e^{i\boldsymbol{k} \cdot (\boldsymbol{x}_e - \boldsymbol{z}_e)} \right) \boldsymbol{\psi}_{e,\boldsymbol{k}} \, .
\end{aligned}
$$

corresponding to the matrix representation $\underline{S}_{h,\boldsymbol{k}} = \left(\underline{S}_{e,v} \right)_{e \in \mathcal{E}_h, v \in \mathcal{V}_h}$ with

$$S_{h,\boldsymbol{k}} = \sum_{e \in \mathcal{E}_h, \, v \in \mathcal{V}_h} \underline{S}_{e,v} \phi'_{v,\boldsymbol{k}} \otimes \boldsymbol{\psi}_{e,\boldsymbol{k}} \, , \quad \underline{S}_{e,v} = \phi_{v,\boldsymbol{k}}(\boldsymbol{y}_e) e^{i\boldsymbol{k} \cdot (\boldsymbol{y}_e - \boldsymbol{z}_e)} - \phi_{v,\boldsymbol{k}}(\boldsymbol{x}_e) e^{i\boldsymbol{k} \cdot (\boldsymbol{x}_e - \boldsymbol{z}_e)}.$$

Moreover, this operator satisfies the compatibility condition

$$b_{\boldsymbol{k}}(S_{h,\boldsymbol{k}} p_{h,\boldsymbol{k}}, q_{h,\boldsymbol{k}}) = c_{\boldsymbol{k}}(p_{h,\boldsymbol{k}}, q_{h,\boldsymbol{k}}) \, ,$$

i. e., the operator equation $B_{h,\boldsymbol{k}} S_{h,\boldsymbol{k}} = C_{h,\boldsymbol{k}}$.

A direct consequence of this compatibility is the inf-sup stability of the finite element pair $\boldsymbol{V}_{h,\boldsymbol{k}} \times \hat{Q}_{h,\boldsymbol{k}}$, i. e.,

$$
\begin{aligned}
\sup_{\boldsymbol{v}_{h,\boldsymbol{k}} \in \boldsymbol{V}_{h,\boldsymbol{k}} \setminus \{0\}} \frac{b_{\boldsymbol{k}}(\boldsymbol{v}_{h,\boldsymbol{k}}, q_{h,\boldsymbol{k}})}{\|\boldsymbol{v}_{h,\boldsymbol{k}}\|_{H(\mathrm{curl},\Omega)}} &\geq \frac{b_{\boldsymbol{k}}(S_{h,\boldsymbol{k}} q_{h,\boldsymbol{k}}, q_{h,\boldsymbol{k}})}{\|S_{h,\boldsymbol{k}} q_{h,\boldsymbol{k}}\|_{H(\mathrm{curl},\Omega)}} = \frac{c_{\boldsymbol{k}}(q_{h,\boldsymbol{k}}, q_{h,\boldsymbol{k}})}{\|S_{h,\boldsymbol{k}} q_{h,\boldsymbol{k}}\|_{L^2(\Omega,\mathbb{C}^3)}} \\
&\geq \beta \, \|q_{h,\boldsymbol{k}}\|_{H^1(\Omega,\mathbb{C}^3)}
\end{aligned}
\tag{2.25}
$$

for all $q_{h,\boldsymbol{k}} \in \hat{Q}_{h,\boldsymbol{k}}$, where the constant $\beta > 0$ is independent of the mesh size and can be computed from the Poincaré–Friedrich constant. Recall how this is used to prove Theorem 2.1.8.

Further compatibility properties can be formulated as an exact sequence property, cf. [26, 19]. In particular, the operator $S_{h,\boldsymbol{k}}$ maps into the kernel of $A_{h,\boldsymbol{k}}$, i. e., $A_{h,\boldsymbol{k}} S_{h,\boldsymbol{k}} = 0$. In order to obtain a preconditioner for the LOBPCG method, we define the stabilised Maxwell operator $A_{h,\boldsymbol{k}}^{\delta} = A_{h,\boldsymbol{k}} + \delta M_{h,\boldsymbol{k}}$ for $\delta > 0$. This operator is self-adjoint and regular, and we will see below how to construct a robust multigrid preconditioner for this operator.

2.2.6 The projection onto the constrained space

Since for the constrained space $\boldsymbol{X}_{h,\boldsymbol{k}}$ no local basis exists, we construct approximations in the full space $\boldsymbol{V}_{h,\boldsymbol{k}}$ and we provide a projection onto $\boldsymbol{X}_{h,\boldsymbol{k}}$.

This projection can be constructed as follows: for $\boldsymbol{v}_{h,\boldsymbol{k}} \in \boldsymbol{V}_{h,\boldsymbol{k}}$ we construct $p_{h,\boldsymbol{k}} \in Q_{h,\boldsymbol{k}}$ such that $\boldsymbol{v}_{h,\boldsymbol{k}} - \boldsymbol{\nabla}_{\boldsymbol{k}} p_{h,\boldsymbol{k}} \in \boldsymbol{X}_{h,\boldsymbol{k}}$, i. e.,

$$
\begin{aligned}
0 &= b_{\boldsymbol{k}}(\boldsymbol{v}_{h,\boldsymbol{k}} - \boldsymbol{\nabla}_{\boldsymbol{k}} p_{h,\boldsymbol{k}}, q_{h,\boldsymbol{k}}) \\
&= b_{\boldsymbol{k}}(\boldsymbol{v}_{h,\boldsymbol{k}}, q_{h,\boldsymbol{k}}) - c_{\boldsymbol{k}}(p_{h,\boldsymbol{k}}, q_{h,\boldsymbol{k}}) \quad \text{for all} \quad q_{h,\boldsymbol{k}} \in Q_{h,\boldsymbol{k}}.
\end{aligned}
$$

Thus, we have $B_{h,\boldsymbol{k}} \boldsymbol{v}_{h,\boldsymbol{k}} = C_{h,\boldsymbol{k}} p_{h,\boldsymbol{k}}$ and $\boldsymbol{v}_{h,\boldsymbol{k}} = S_{h,\boldsymbol{k}} p_{h,\boldsymbol{k}}$. Since the kernels of $S_{h,\boldsymbol{k}}$ and $C_{h,\boldsymbol{k}}$ coincide, the identity $S_{h,\boldsymbol{k}} p_{h,\boldsymbol{k}} = S_{h,\boldsymbol{k}} \hat{C}_{h,\boldsymbol{k}}^{-1} C_{h,\boldsymbol{k}} p_{h,\boldsymbol{k}}$ holds. Together, this defines a projection operator

$$
P_{h,\boldsymbol{k}} = \text{id} - S_{h,\boldsymbol{k}} \circ \hat{C}_{h,\boldsymbol{k}}^{-1} \circ B_{h,\boldsymbol{k}} \colon \boldsymbol{V}_{h,\boldsymbol{k}} \longrightarrow \boldsymbol{X}_{h,\boldsymbol{k}},
$$

where $\hat{C}_{h,\boldsymbol{k}}^{-1} \colon Q'_{h,\boldsymbol{k}} \longrightarrow \hat{Q}_{h,\boldsymbol{k}}$ gives the solution to a Poisson problem.

In the algorithms below for the eigenvalue computation it will be required that the evaluation of the projection is quite accurate. More precisely, we use an iterative scheme which reduces the initial residual norm by a suitable factor, so that the algebraic error of the approximate projection is considerably smaller than the discretisation error.

2.2.7 Operator properties and finite element convergence

The discretisation $\boldsymbol{X}_{h,\boldsymbol{k}}$ is non-conforming, i. e., $\boldsymbol{X}_{h,\boldsymbol{k}} \not\subset \boldsymbol{X}$. The consistency error of the non-conformity is controlled by the inf-sup stability (2.25), the well-posedness of the inverse follows from the discrete ellipticity (2.12).

The construction of the phase-shifted spaces allows to apply the arguments of Chapter 2.1 (that relate to the case $\boldsymbol{k} = \boldsymbol{0}$) to the actual setting. Especially we have refer to Theorem 2.1.30 for convergence of eigenvalues and eigenfunctions. This method has been developed in [4].

2.2.8 The preconditioned inverse iteration including projections

The inverse iteration is the most simple algorithm for the computation of the smallest eigenvalue and the corresponding eigenmode in the space $\boldsymbol{V}_{h,\boldsymbol{k}}$ of divergence-free functions. Unfortunately, this requires to solve in every step a Maxwell problem in $\boldsymbol{V}_{h,\boldsymbol{k}}$.

In order to get rid of the exact Maxwell solution in every step, we assume that it is sufficient to apply a (linear) preconditioner of the Maxwell problem. Since the Maxwell operator has a large kernel, we consider the shifted operator

$$
A_{h,\boldsymbol{k}}^{\delta} = A_{h,\boldsymbol{k}} + \delta M_{h,\boldsymbol{k}} \colon \boldsymbol{V}_{h,\boldsymbol{k}} \longrightarrow \boldsymbol{V}'_{h,\boldsymbol{k}}
$$

and the corresponding sesqui-linear form (2.22) for some fixed operator shift $\delta > 0$, and we use in the iteration a preconditioner

$$
T_{h,\boldsymbol{k}} \colon \boldsymbol{V}'_{h,\boldsymbol{k}} \longrightarrow \boldsymbol{V}_{h,\boldsymbol{k}}
$$

for $A_{h,\boldsymbol{k}}^{\delta}$. We require that the preconditioner is optimal in the sense that $T_{h,\boldsymbol{k}}$ is spectrally equivalent to the inverse of $A_{h,\boldsymbol{k}}^{\delta}$, i.e,

$$C_0 \, a_{\boldsymbol{k}}^{\delta}(\boldsymbol{v}_{h,\boldsymbol{k}}, \boldsymbol{v}_{h,\boldsymbol{k}}) \leq a_{\boldsymbol{k}}^{\delta}(T_{h,\boldsymbol{k}} A_{h,\boldsymbol{k}}^{\delta} \boldsymbol{v}_{h,\boldsymbol{k}}, \boldsymbol{v}_{h,\boldsymbol{k}}) \leq C_1 \, a_{\boldsymbol{k}}^{\delta}(\boldsymbol{v}_{h,\boldsymbol{k}}, \boldsymbol{v}_{h,\boldsymbol{k}}) \qquad (2.26)$$

for all $\boldsymbol{v}_{h,\boldsymbol{k}} \in \boldsymbol{V}_{h,\boldsymbol{k}}$ with constants $C_1 > C_0 > 0$ independent of the mesh size h and the operator shift δ.

It turns out that it is not required to include the divergence constraint into the preconditioner $T_{h,\boldsymbol{k}}$; it is sufficient to include a projection step onto the constrained space $\boldsymbol{X}_{h,\boldsymbol{k}}$ in every iteration. Together, this results in algorithm 1 for the projected inverse iteration.

Algorithm 1 Preconditioned inverse iteration.

(S0) Choose randomly $\boldsymbol{v}_{h,\boldsymbol{k}} \in \boldsymbol{V}_{h,\boldsymbol{k}} \setminus \{0\}$.
(S1) Compute the projection $\boldsymbol{w}_{h,\boldsymbol{k}} = P_{h,\boldsymbol{k}} \boldsymbol{v}_{h,\boldsymbol{k}} \in \boldsymbol{X}_{h,\boldsymbol{k}}$.
(S2) Compute a normalisation

$$\boldsymbol{u}_{h,\boldsymbol{k}} = \frac{1}{\sqrt{m(\boldsymbol{w}_{h,\boldsymbol{k}}, \boldsymbol{w}_{h,\boldsymbol{k}})}} \, \boldsymbol{w}_{h,\boldsymbol{k}} \, .$$

(S3) Compute the eigenvalue approximation

$$\lambda_{h,\boldsymbol{k}} = a_{\boldsymbol{k}}(\boldsymbol{u}_{h,\boldsymbol{k}}, \boldsymbol{u}_{h,\boldsymbol{k}})$$

and the residual

$$\boldsymbol{r}_{h,\boldsymbol{k}} \;=\; A_{h,\boldsymbol{k}} \boldsymbol{u}_{h,\boldsymbol{k}} - \lambda_{h,\boldsymbol{k}} M_{h,\boldsymbol{k}} \boldsymbol{u}_{h,\boldsymbol{k}} \in \boldsymbol{X}'_{h,\boldsymbol{k}} \, .$$

(S4) If the residual norm $\|\boldsymbol{r}_{h,\boldsymbol{k}}\|$ is small enough, then STOP.
(S5) Apply the preconditioner: Compute $\boldsymbol{v}_{h,\boldsymbol{k}} := T_{h,\boldsymbol{k}} \boldsymbol{r}_{h,\boldsymbol{k}}$. Go to step (S1).

This method is proposed and analysed in [20], where more details are given. It is observed that the spectral estimate (2.26) transfers to the combined method which includes the projection, i. e. , the condition number of the combined operator $P_{h,\boldsymbol{k}} T_{h,\boldsymbol{k}} A_{h,\boldsymbol{k}}$ (restricted to the constrained space $\boldsymbol{X}_{h,\boldsymbol{k}}$) is bounded independently of the mesh size and the shift parameter.

It turns out that one or two preconditioning steps in (S5) are sufficient. On the other hand, for the evaluation of the projection the approximation by simple preconditioning is not enough (although an exact projection is not required): if the projection is approximated only roughly, the iteration will converge to a vector in the kernel of $A_{h,\boldsymbol{k}}$, i. e., to an eigenvector with eigenvalue 0.

The convergence of this algorithm is linear, provided that the first eigenvalue is isolated. The convergence rate depends on the quotient between first and second eigenvalue. In principle, this can be improved if the operator shift δ is adapted in every iteration step. The optimal choice $\delta = -\lambda_{h,\boldsymbol{k}}$ with the approximated eigenvalue of step (S3) leads to cubic convergence of the eigenvalue iteration, but it requires to solve nearly indefinite linear problems, so that the iteration with simple preconditioning in (S5) is far more efficient.

2.2.9 The Ritz–Galerkin projection

Since in our application we cannot guarantee that the first eigenvalue is simple, and since we aim for the simultaneous approximation of several eigenvalues in order to identify a band gap, we need a block version of the inverse iteration.

Let $N > 0$ be a small number, e. g., $N = 10$. For a given set $\{u_{h,k}^1, \ldots, u_{h,k}^N\} \subset X_{h,k}$ of linear independent functions we can define the Galerkin approximation of the eigenvalue problem (2.23) in the subspace $X_{h,k}^N = \text{Span}\{u_{h,k}^1, \ldots, u_{h,k}^N\}$: Set up the Hermitian matrices

$$A_N = \left(a_k(u_{h,k}^l, u_{h,k}^n) \right)_{l,n=1,\ldots,N}, \quad M_N = \left(m(u_{h,k}^l, u_{h,k}^n) \right)_{l,n=1,\ldots,N} \in \mathbb{C}^{N \times N}$$

and solve a matrix eigenvalue problem: find eigenvectors $z^n \in \mathbb{C}^N \setminus \{0\}$ and eigenvalues $\mu_n \in \mathbb{R}$ with

$$A_N z^n = \mu_n M_N z^n, \qquad n = 1, \ldots, N, \tag{2.27}$$

where we assume that the eigenvalues are ordered such that $\mu_1 \leq \mu_2 \leq \cdots \leq \mu_N$. We set $(z^1, \ldots, z^N) = \mathcal{Z}(u_{h,k}^1, \ldots, u_{h,k}^N)$ for this procedure. Then, compute the Ritz–Galerkin projection

$$y_{h,k}^n = \sum_{l=1}^N z_l^n u_{h,k}^l \in X_{h,k}, \qquad n = 1, \ldots, N. \tag{2.28}$$

By construction, we have for $l \neq n$

$$a_k(y_{h,k}^l, y_{h,k}^n) = 0, \qquad m(y_{h,k}^l, y_{h,k}^n) = 0,$$

and for the Rayleigh quotient

$$\mu_n = \frac{a_k(y_{h,k}^n, y_{h,k}^n)}{m(y_{h,k}^n, y_{h,k}^n)}.$$

The preconditioned inverse iteration can be started simultaneously for the N vectors $u_{h,k}^n$, if in every step a Ritz–Galerkin projection is included. Again, we expect linear convergence of the eigenvalue approximations $\mu_1, \ldots, \mu_n$ to $\lambda_{h,k}^1 \leq \cdots \leq \lambda_{h,k}^n$, provided that for some $1 < n \leq N$ a spectral gap $\lambda_{h,k}^{n+1} - \lambda_{h,k}^n > 0$ exists. The convergence factor (in case of exact projections and suitable start vectors) can be estimated by

$$q = 1 - \frac{2}{1 + C_1/C_0} \left(1 - \frac{\lambda_{h,k}^n + \delta}{\lambda_{h,k}^{n+1} + \delta} \right), \tag{2.29}$$

cf. [25, Thm. 9].

2.2.10 The modified LOBPCG method including projections

The LOBPCG method is introduced in [24] in order to improve the convergence of the preconditioned inverse iteration by augmenting the subspace of the Ritz–Galerkin projection. The main idea is inspired by the tree-term recursion of Lanczos method: the Ritz–Galerkin subspace is built of three components, the actual approximations $y_{h,k}^n$, the preconditioned residuals $w_{h,k}^n$, and suitable conjugated directions $d_{h,k}^n$.

Here, we consider the modification introduced by [30] where the preconditioned projected inverse iteration [20] is combined with the LOBPCG method [24], cf. Algorithm 2.

It turns out that this algorithm is far more robust than a simple block version of the preconditioned projected inverse iteration. Nevertheless, a careful choice of stopping criteria is required. In our implementation, we use the following strategies:

(1) In every Ritz–Galerkin step it should be checked with a Gram–Schmidt orthogonalisation, whether the vectors $u_{h,k}^1, \ldots, u_{h,k}^N$ spanning the Ritz–Galerkin subspace are linearly independent: for $n = 2, \ldots, N$ compute

$$v_{h,k}^n = u_{h,k}^n - \sum_{l=1}^{n-1} m(u_{h,k}^l, u_{h,k}^n)\, u_{h,k}^l$$

so that $v_{h,k}^n$ is orthogonal to $u_{h,k}^1, \ldots, u_{h,k}^{n-1}$. If $\|v_{h,k}^n\| = \sqrt{m(v_{h,k}^n, v_{h,k}^n)} \geq \epsilon_1$, set $u_{h,k}^n = v_{h,k}^n / \|v_{h,k}^n\|$. Otherwise, the vector $v_{h,k}^n$ is nearly linearly depending from $u_{h,k}^1, \ldots, u_{h,k}^{n-1}$. Then, $u_{h,k}^n$ is replaced by a new random vector and the orthogonalisation procedure is repeated.

Here, $\epsilon_1 = 10^{-8}$ is appropriate, if the stopping criterion for the residual is not too small.

(2) The projections $w_{h,k}^n = P_{h,k} v_{h,k}^n$ in (S0) and (S2) are computed only approximately: we compute $q_{h,k} \in Q_{h,k}$ such that $\|C_{h,k} q_{h,k} - B_{h,k} v_{h,k}^n\| \leq \epsilon_2 + \theta_2 \|B_{h,k} v_{h,k}^n\|$ and set $w_{h,k}^n = v_{h,k}^n - S_{h,k} q_{h,k} \in V_{h,k}$, i.e., in general $w_{h,k}^n \notin X_{h,k}$.

Here, $\epsilon_2 = 10^{-8}$ and a residual reduction $\theta_2 = 10^{-7}$ is appropriate. For simplicity, one can use a simple Euclidean vector norm since the strong residual reduction already ensures that the relative error is small enough.

Note that the eigenvalue accuracy which can be obtained with the projected LOBPCG method is strictly limited by the accuracy of the projection.

(3) On the other hand, the algorithm does not require an accurate solution of the preconditioning step in (S2). In our implementation it is realised as follows: we apply a few iterations of a Krylov method (preconditioned with $T_{h,k}$) such that $\|A_{h,k}^\delta v_{h,k}^n - r_{h,k}^n\| \leq \epsilon_3 + \theta_3 \|r_{h,k}^n\|$. Moreover, we observe that the algorithm is not sensitive with respect to the operator shift $\delta > 0$ (in

Algorithm 2 The modified LOPCG method

(S0) Choose random vectors $\boldsymbol{v}_{h,\boldsymbol{k}}^1, \ldots, \boldsymbol{v}_{h,\boldsymbol{k}}^N \in \boldsymbol{V}_{h,\boldsymbol{k}}$.
For $n = 1, \ldots, N$, compute the projections and normalisations

$$\boldsymbol{w}_{h,\boldsymbol{k}}^n = P_{h,\boldsymbol{k}} \boldsymbol{v}_{h,\boldsymbol{k}}^n \in \boldsymbol{X}_{h,\boldsymbol{k}}, \qquad u_{h,\boldsymbol{k}} = \frac{1}{\sqrt{m(\boldsymbol{w}_{h,\boldsymbol{k}}, \boldsymbol{w}_{h,\boldsymbol{k}})}} \, \boldsymbol{w}_{h,\boldsymbol{k}}.$$

Then, compute the Ritz–Galerkin eigenvalues $\lambda^1, \ldots, \lambda^N$ and eigenvectors

$$(z^1, \ldots, z^N) = \mathcal{Z}(\boldsymbol{u}_{h,\boldsymbol{k}}^1, \ldots, \boldsymbol{u}_{h,\boldsymbol{k}}^N)$$

and the Ritz–Galerkin projection

$$\boldsymbol{y}_{h,\boldsymbol{k}}^n = \sum_{m=1}^N z_m^n \boldsymbol{u}_{h,\boldsymbol{k}}^m \in \boldsymbol{X}_{h,\boldsymbol{k}}, \qquad n = 1, \ldots, N.$$

(S1) For $n = 1, \ldots, N$, compute the residuals

$$\boldsymbol{r}_{h,\boldsymbol{k}}^n = A_{h,\boldsymbol{k}} \boldsymbol{y}_{h,\boldsymbol{k}}^n - \lambda^n M_{h,\boldsymbol{k}} \boldsymbol{y}_{h,\boldsymbol{k}}^n \in \boldsymbol{V}_{h,\boldsymbol{k}}',$$

and check for convergence.

(S2) For $n = 1, \ldots, N$, apply the preconditioner $\boldsymbol{v}_{h,\boldsymbol{k}}^n = T_{h,\boldsymbol{k}} \boldsymbol{r}_{h,\boldsymbol{k}}^n \in \boldsymbol{V}_{h,\boldsymbol{k}}$ and compute the projection $\boldsymbol{w}_{h,\boldsymbol{k}}^n := P_{h,\boldsymbol{k}} \boldsymbol{v}_{h,\boldsymbol{k}}^n \in \boldsymbol{X}_{h,\boldsymbol{k}}$.

(S3) Perform this step in the first iteration; otherwise go to (S4).
Compute the Ritz–Galerkin eigenvectors

$$(z^1, \ldots, z^{2N}) = \mathcal{Z}(\boldsymbol{w}_{h,\boldsymbol{k}}^1, \ldots, \boldsymbol{w}_{h,\boldsymbol{k}}^N, \boldsymbol{y}_{h,\boldsymbol{k}}^1, \ldots, \boldsymbol{y}_{h,\boldsymbol{k}}^N).$$

For $n = 1, \ldots, N$, set

$$\begin{aligned}
\boldsymbol{d}_{h,\boldsymbol{k}}^n &= \sum_{l=1}^N z_l^n \boldsymbol{w}_{h,\boldsymbol{k}}^l \in \boldsymbol{X}_{h,\boldsymbol{k}}, \\
\tilde{\boldsymbol{y}}_{h,\boldsymbol{k}}^n &= \boldsymbol{d}_{h,\boldsymbol{k}}^n + \sum_{l=1}^N z_{N+l}^n \boldsymbol{y}_{h,\boldsymbol{k}}^l \in \boldsymbol{X}_{h,\boldsymbol{k}}.
\end{aligned}$$

Then, set $\boldsymbol{y}_{h,\boldsymbol{k}}^n = \tilde{\boldsymbol{y}}_{h,\boldsymbol{k}}^n$ for $n = 1, \ldots, N$. Go to (S1).

(S4) Compute the Ritz–Galerkin eigenvectors

$$(z^1, \ldots, z^{3N}) = \mathcal{Z}(\boldsymbol{w}_{h,\boldsymbol{k}}^1, \ldots, \boldsymbol{w}_{h,\boldsymbol{k}}^N, \boldsymbol{y}_{h,\boldsymbol{k}}^1, \ldots, \boldsymbol{y}_{h,\boldsymbol{k}}^N, \boldsymbol{d}_{h,\boldsymbol{k}}^1, \ldots, \boldsymbol{d}_{h,\boldsymbol{k}}^N).$$

For $n = 1, \ldots, N$, set

$$\begin{aligned}
\tilde{\boldsymbol{d}}_{h,\boldsymbol{k}}^n &= \sum_{l=1}^N z_l^n \boldsymbol{w}_{h,\boldsymbol{k}}^l + z_{2N+l}^n \boldsymbol{d}_{h,\boldsymbol{k}}^l, \\
\tilde{\boldsymbol{y}}_{h,\boldsymbol{k}}^n &= \tilde{\boldsymbol{d}}_{h,\boldsymbol{k}}^n + \sum_{l=1}^N z_{N+l}^n \boldsymbol{y}_{h,\boldsymbol{k}}^l.
\end{aligned}$$

Then, set $\boldsymbol{y}_{h,\boldsymbol{k}}^n = \tilde{\boldsymbol{y}}_{h,\boldsymbol{k}}^n$ and $\boldsymbol{d}_{h,\boldsymbol{k}}^n = \tilde{\boldsymbol{d}}_{h,\boldsymbol{k}}^n$ for $n = 1, \ldots, N$. Go to (S1).

particular, $A_{h,\boldsymbol{k}}^{\delta}$ and $A_{h,\boldsymbol{k}}$ have the same eigenvectors). From (2.29) we can estimate that a residual reduction $\theta_3 = 0.1$ and an operator shift $\delta = 0.1\lambda_{\boldsymbol{k},h}^1$ is a suitable choice (indeed, a further reduction to a smaller $\delta > 0$ or a smaller θ_3 does not reduce the required number of LOBPCG steps).

(4) After a few LOBPCG iteration steps the smallest eigenvalues are already very accurate (if N is large enough). Depending on the accuracy of the projection it may be not possible to improve the corresponding eigenmode. So in step (S1) we determine $0 \leq n_{\mathrm{conv}} \leq N$ such that $\|\boldsymbol{r}_{h,\boldsymbol{k}}^n\| \leq \epsilon_4$ for $n = 1, \ldots, n_{\mathrm{conv}}$. Then, we proceed the method with the vectors

$$\boldsymbol{w}_{h,\boldsymbol{k}}^{n_{\mathrm{conv}}+1}, \ldots, \boldsymbol{w}_{h,\boldsymbol{k}}^N, \boldsymbol{y}_{h,\boldsymbol{k}}^{n_{\mathrm{conv}}+1}, \ldots, \boldsymbol{y}_{h,\boldsymbol{k}}^N, \boldsymbol{d}_{h,\boldsymbol{k}}^{n_{\mathrm{conv}}+1}, \ldots, \boldsymbol{d}_{h,\boldsymbol{k}}^N$$

in step (S4).

The overall procedure is more efficient if N is larger than the desired number of eigenvalues. Thus, one should stop the LOBPCG iteration if $n_{\mathrm{conv}} \geq N/2$. Here, $\epsilon_4 \approx \sqrt{\epsilon_2}$ is appropriate. Then, we can expect a relative accuracy of $\epsilon_4^2 \approx \epsilon_2$ for the eigenvalues. Since in this stopping criterion the residual norm is used for estimating the error, a mesh-independent evaluation is required. This is obtained, e. g., by the application of the preconditioner: set

$$\|\boldsymbol{r}_{h,\boldsymbol{k}}^n\| = \sqrt{\langle \boldsymbol{r}_{h,\boldsymbol{k}}^n, \boldsymbol{w}_{h,\boldsymbol{k}}^n \rangle} = \sqrt{\langle \boldsymbol{r}_{h,\boldsymbol{k}}^n, T_{\boldsymbol{k},\boldsymbol{k}} \boldsymbol{r}_{h,\boldsymbol{k}}^n \rangle} \, .$$

A simple alternative (which avoids to apply the preconditioner) is to use one sweep of a Gauss–Seidel relaxation $R_{j,\boldsymbol{k}} \colon \boldsymbol{V}_{j,\boldsymbol{k}}' \longrightarrow \boldsymbol{V}_{j,\boldsymbol{k}}$ for $A_{j,\boldsymbol{k}}$, i. e., to use $\|\boldsymbol{r}_{h,\boldsymbol{k}}^n\| = \sqrt{\langle \boldsymbol{r}_{h,\boldsymbol{k}}^n, R_{\boldsymbol{k},\boldsymbol{k}} \boldsymbol{r}_{h,\boldsymbol{k}}^n \rangle}$.

More details on different variants of the LOBPCG method and the suitable choice of parameters are discussed in [9].

2.2.11   A multigrid preconditioner for the Maxwell operator

A suitable multigrid preconditioner $T_{j,\boldsymbol{k}} \colon \boldsymbol{V}_{j,\boldsymbol{k}}' \longrightarrow \boldsymbol{V}_{j,\boldsymbol{k}}$ satisfying (2.26) with $C_1 > C_0 > 0$ independent of the mesh size and robust for $0 < \delta \leq \delta_0$ is introduced by Hiptmair [18]. Therefore, let $\boldsymbol{V}_{0,\boldsymbol{k}} \subset \boldsymbol{V}_{1,\boldsymbol{k}} \subset \cdots \subset \boldsymbol{V}_{J,\boldsymbol{k}} \subset \boldsymbol{V}$ be a nested sequence of curl-conforming finite element spaces with phase-shifted basis and with mesh size $h_j = 2^{-j} h_0$ on level $j = 0, \ldots, J$, and let $Q_{0,\boldsymbol{k}} \subset Q_{1,\boldsymbol{k}} \subset \cdots \subset Q_{J,\boldsymbol{k}} \subset Q$ be the corresponding H^1-conforming spaces with phase-shifted basis.

The main observation is that a stable splitting corresponding to the decomposition

$$\boldsymbol{V}_{J,\boldsymbol{k}} = \boldsymbol{V}_{0,\boldsymbol{k}} + \sum_{j=1}^{J} \boldsymbol{V}_{j,\boldsymbol{k}} + \sum_{j=1}^{J} \boldsymbol{\nabla}_{\boldsymbol{k}} Q_{j,\boldsymbol{k}}$$

with respect to the energy norm of $A_{J,\boldsymbol{k}}^{\delta}$ exists. As a consequence we obtain multi-grid convergence for a hybrid smoother which alternatively operates on $\boldsymbol{V}_{j,\boldsymbol{k}}$ and

$\nabla_k Q_{j,k}$. The multigrid preconditioner is defined recursively: for $j = 0$, define $T_{0,k} = \left(A_{0,k}^\delta\right)^{-1}$. For $j > 0$, the definition of $T_{j,k}$ requires:

1) a prolongation operator $I_{j,k}\colon V_{j-1,k} \longrightarrow V_{j,k}$;

2) the adjoint operator (restriction operator) $I'_{j,k}\colon V'_{j,k} \longrightarrow V'_{j-1,k}$;

3) a smoother $R_{j,k}\colon V'_{j,k} \longrightarrow V_{j,k}$ for $A_{j,k}$;

4) a smoother $D_{j,k}\colon Q'_{j,k} \longrightarrow Q_{j,k}$ for $C_{j,k}$;

5) a transfer operator $S_{j,k}\colon Q_{j,k} \longrightarrow V_{j,k}$;

6) the adjoint transfer operator $S'_{j,k}\colon V'_{j,k} \longrightarrow Q'_{j,k}$.

Now, define $T_{j,k}$ by

$$
\begin{aligned}
\mathrm{id} &- A_{j,k}^\delta T_{j,k} \\
&= \left(\mathrm{id} - A_{j,k}^\delta I_{j,k} T_{j-1,k} I'_{j,k}\right)\left(\mathrm{id} - \delta^{-1} A_{j,k}^\delta S_{j,k} D_{j,k} S'_{j,k}\right)^\nu \left(\mathrm{id} - A_{j,k}^\delta R_{j,k}\right)^\mu.
\end{aligned}
$$

For given residual $r_{J,k}$, the result $v_{J,k} = T_{J,k} r_{J,k}$ is computed by Algorithm 3.

Algorithm 3 Multigrid preconditioner $v_{J,k} = T_{J,k} r_{J,k}$ with hybrid smoother.

(S0) Set $j = J$.
(S1) Set $v_{j,k} = 0$. For $\kappa = 1, \ldots, \mu$ compute

$$
\begin{aligned}
w_{j,k} &= R_{j,k} r_{j,k}, \\
v_{j,k} &:= v_{j,k} + w_{j,k}, \\
r_{j,k} &:= r_{j,k} - A_{j,k}^\delta w_{j,k}.
\end{aligned}
$$

(S2) For $\kappa = 1, \ldots, \nu$ compute

$$
\begin{aligned}
w_{j,k} &= \delta^{-1} A_{j,k}^\delta S_{j,k} D_{j,k} S'_{j,k} r_{j,k}, \\
v_{j,k} &:= v_{j,k} + w_{j,k}, \\
r_{j,k} &:= r_{j,k} - A_{j,k}^\delta w_{j,k}.
\end{aligned}
$$

(S3) If $j > 0$, set $r_{j-1,k} = I'_{j,k} r_{j,k}$, $j := j - 1$, and go to (S1).
(S4) If $j = 0$, set $v_{0,k} := T_{0,k} r_{0,k}$.
(S5) For $j = 1, \ldots, J$ set

$$
v_{j,k} := v_{j,k} + I_{j,k} v_{j-1,k}.
$$

Return the result $v_{J,k}$.

The prolongation is defined by the simple embedding $V_{j-1,k} \subset V_{j,k}$, i.e.,

$$
I_{j,k} v_{j-1,k} = \sum_{e \in \mathcal{E}_j} \langle \psi'_{e,k}, v_{j-1,k} \rangle \psi_{e,k},
$$

and it is represented by the matrix $\underline{I}_{j,k} = \left(\underline{I}_{e_1,e_2}\right)_{(e_1,e_2)\in\mathcal{E}_j\times\mathcal{E}_{j-1}}$

$$I_{j,k} = \sum_{(e_1,e_2)\in\mathcal{E}_j\times\mathcal{E}_{j-1}} \underline{I}_{e_1,e_2}\,\psi'_{e_2,k}\otimes\psi_{e_1,k}, \qquad \underline{I}_{e_1,e_2} = \langle\psi'_{e_1,k},\psi_{e_2,k}\rangle.$$

The restriction is represented by the transposed matrix $\underline{I}^T_{j,k}$, i.e.,

$$I'_{j,k} = \sum_{(e_2,e_1)\in\mathcal{E}_{j-1}\times\mathcal{E}_j} \underline{I}_{e_1,e_2}\,\psi_{e_1,k}\otimes\psi'_{e_2,k}.$$

Note that this implies that the transfer operators depend on the phase-shift, but again (as it is observed in Chapter 2.2.4) by a simple scaling the operators can be constructed from the standard operators for $\mathbf{k}=\mathbf{0}$.

The smoothing operators are constructed algebraically from the matrix representations of $A_{j,k}$ and $C_{j,k}$. E.g., the Gauss–Seidel relaxation is defined by

$$R_{j,k} = \sum_{e_1,e_2\in\mathcal{E}_j} \underline{R}_{e_1,e_2}\,\psi_{e_2,k}\otimes\psi_{e_1,k}, \qquad \underline{R}_{j,k} = \left(\operatorname{diag}\underline{A}_{j,k} + \operatorname{lower}\underline{A}_{j,k}\right)^{-1},$$

$$D_{h,k} = \sum_{v_1,v_2\in\mathcal{V}_h} \underline{D}_{v_1,v_2}\,\phi_{v_2,k}\otimes\phi_{v_1,k}, \qquad \underline{D}_{j,k} = \left(\operatorname{diag}\underline{C}_{j,k} + \operatorname{lower}\underline{D}_{j,k}\right)^{-1}.$$

Note that Gauss–Seidel smoothing depends on the numbering of the indices.

For the parallel smoothers $R_{j,k}$ and $D_{j,k}$ suitable damping is required; in addition, we use multiple smoothing $\nu,\mu>1$, and we use a symmetric variant of the presented algorithm.

In the same way, we construct the Laplace preconditioner $U_{j,k}: Q'_{j,k}\longrightarrow Q_{j,k}$ within the projection step. Again we set $U_{0,k}=C^{-1}_{0,k}$, and recursively $U_{j,k}$ is defined for $j>0$ by

$$\operatorname{id} - C_{j,k}U_{j,k} = \left(\operatorname{id} - C_{j,k}J_{j,k}U_{j-1,k}J'_{j,k}\right)\left(\operatorname{id} - C_{j,k}D_{j-1,k}\right)^{\nu},$$

where $J_{j,k}: Q_{j-1,k}\longrightarrow Q_{j,k}$ and $J'_{j,k}: Q'_{j,k}\longrightarrow Q'_{j-1,k}$ are the prolongation and restriction operators, respectively (again obtained from the standard transfer operators by a scaling with respect to the phase-shift).

2.2.12 Numerical results for a band gap computation

Finally, we summarise results from [9] for the eigenvalue computation of a specific configuration. The computation is realised in the parallel finite element code M++ [28] which supports fully parallel multigrid methods [29].

We consider a material distribution proposed in [16], where we use a material with permittivity $\varepsilon_r = 13$ (which is in a realistic range for, e.g., silicon), and for the empty space we have $\varepsilon_r = 1$. The distribution is a layered structure, cf. Fig. 2.1. The structure is highly symmetric in xy plane, but is not symmetric in z direction. The configuration consists of silicon blocks, the block thickness in the periodic structure is 0.25. We start in level 0 with one hexahedron, then we use regular refinement, see Tab. 2.1. Thus, level 3 is the minimal level where the material distribution is represented exactly, i.e., the distributions is aligned with the cells.

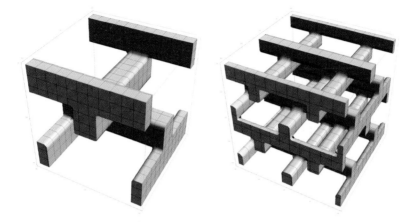

Figure 2.1: A layered structure in the periodicity cell $[0, 1]^3$ (left) and (for illustration) in $[0, 2]^3$ showing 8 periodicity cells (right).

level j	number of cells	d.o.f. in $\boldsymbol{X}_h$	d.o.f. in Q_h
2	64		
3	512	1 944	729
4	4 096	13 872	4 913
5	32 768	104 544	35 937
6	262 144	811 200	274 625
7	2 097 152	6 390 144	2 146 689
8	16 777 216	50 725 632	16 974 593

Table 2.1: Refinement levels, number of cells and degrees of freedom for the test configuration.

We test the convergence properties of the LOBPCG method for a fixed parameter $\boldsymbol{k} = (3, 1, -2)$, where we use $\epsilon_4 = 10^{-4}$ as stopping criterion. The results in Tab. 2.2 indicates that the convergence is almost independent of the mesh size, and that the number of iterations only slowly increases with the number of computed eigenvalues.

All results are obtained on a parallel Linux cluster. Since the multigrid solver is of optimal complexity, we expect an optimal scaling behaviour of the parallel

N #	level 3	level 4	level 5	level 6
1	6	6	6	7
2	6	7	7	7
3	7	7	7	8
4	7	8	8	8
5	9	10	10	11
6	9	10	11	11
7	11	12	12	12
8	11	12	12	12
9	12	13	13	13
10	12	13	14	14

Table 2.2: Number of iterations of the projected LOBPCG method for the convergence of the first N eigenvalues and eigenvectors depending on the refinement level.

algorithm, provided the coarse problem is fine enough. For a fixed-size numerical performance test (cf. Tab. 2.3) we observe good scalability only up to 256 processor kernels. Optimal scalability is obtain for fixed load per processor kernel up to 50 million unknowns, cf. Tab. 2.4.

processor kernels	multigrid aver. conv. rate	computing time	scaling factor
64	0.28	37:47 min.	
			2.12
128	0.28	17:45 min.	
			1.80
256	0.27	9:54 min.	
			1.19
512	0.29	8:20 min.	

Table 2.3: Fixed-size parallel scalability and Maxwell multigrid convergence on refinement level 7.

Next, we study the eigenvalue convergence. Again, we fix the parameter $\boldsymbol{k} = (3, 1, -2)$, and we set $\epsilon_4 = 10^{-4}$ for the defect precision. Then, at least for isolated eigenvalues we may expect algebraic eigenvector convergence $O(\epsilon_4)$ and eigenvalue convergence $O(\epsilon_4^2)$, so that we can assume that the discretisation error is much larger than the algebraic truncation error. The eigenvalue results are presented in Tab. 2.5, and from their asymptotic behaviour we estimate the convergence rate in Tab. 2.6. For smooth eigenfunctions $O(h^2)$ convergence can be expected, but due to the poor regularity resulting from the discontinuous permittivity the observed convergence rate with respect to the mesh size are smaller.

refinement level	d.o.f.	computing time	scaling factor
5	104,544	1:05 min.	
			1.78
6	811,200	1:56 min.	
			5.12
7	6,390,144	9:54 min.	
			8.09
8	50,725,632	80:21 min.	

Table 2.4: Fixed-load parallel performance on 256 processor kernels.

n	level 3	level 4	level 5	level 6	level 7	level 8
1	3.86613	3.81920	3.80119	3.79449	3.79202	3.79111
2	4.13949	4.09004	4.07099	4.06392	4.06132	4.06036
3	5.01881	4.87990	4.82904	4.81021	4.80322	4.80063
4	5.38737	5.24131	5.18723	5.16714	5.15967	5.15690
5	10.19397	9.68279	9.54135	9.50006	9.48747	9.48348
6	10.46622	9.95338	9.80933	9.76693	9.75392	9.74977
7	12.22023	11.26907	11.02740	10.95948	10.93931	10.93307
8	12.35505	11.40819	11.16670	11.09883	11.07870	11.07247
9	13.78493	12.78181	12.51608	12.43920	12.41575	12.40829
10	13.95290	12.93797	12.66950	12.59189	12.56823	12.56071

Table 2.5: Eigenvalue results $\lambda_{j,k}^n$, $n = 1, \ldots, N$, for different refinement levels j.

Finally, we collect the results for different k following a path in the Brillouin zone. The result illustrates the band structure, cf. Fig. 2.3. Although this path does not cover the full Brillouin zone, one expects extrema in the photonic bands on the boundary of the Brillouin zone or at points with high symmetry, so that we can assume that this figure provides the correct band gap information of the given material distribution. Guaranteed band gap computations can be obtained by a finite number of further eigenvalue computations (including upper and lower eigenvalue bounds) and a suitable perturbation argument (cf. [21]).

Note that the eigenfunctions within the Floquet theory are abstract constructions and not physical objects; nevertheless, some illustrations are presented in Fig. 2.4 and 2.5.

n	$\Delta^n(3,4)$		$\Delta^n(4,5)$		$\Delta^n(5,6)$		$\Delta^n(6,7)$		$\Delta^n(7,8)$
1	0.04693	1.38	0.01801	1.43	0.00669	1.44	0.00247	1.44	0.00091
2	0.04945	1.38	0.01905	1.43	0.00707	1.44	0.00260	1.44	0.00096
3	0.13891	1.45	0.05087	1.43	0.01883	1.43	0.00698	1.43	0.00259
4	0.14607	1.43	0.05407	1.43	0.02009	1.43	0.00747	1.43	0.00277
5	0.51118	1.85	0.14144	1.78	0.04130	1.71	0.01259	1.66	0.00399
6	0.51284	1.83	0.14405	1.76	0.04240	1.70	0.01301	1.65	0.00415
7	0.95116	1.98	0.24167	1.83	0.06792	1.75	0.02017	1.69	0.00625
8	0.94685	1.97	0.24149	1.83	0.06787	1.75	0.02013	1.69	0.00623
9	1.00313	1.92	0.26573	1.79	0.07687	1.71	0.02346	1.65	0.00746
10	1.01493	1.92	0.26847	1.79	0.07761	1.71	0.02366	1.65	0.00752

Table 2.6: The eigenvalue convergence rate is estimated by $\Delta^n(j-1,j) = |\lambda^n_{j-1,k} - \lambda^n_{j,k}|$ and the factor $\log_2\left(\Delta^n(j-1,j)/\Delta^n(j,j+1)\right)$.

Bibliography

[1] D. Boffi. Fortin elements and discrete compactness for edge elements. *Numer. Math.*, 87:229–246, 2000.

[2] D. Boffi. Finite element approximations of eigenvalue problems. *Acta Numerica*, 19:1–120, 2010.

[3] D. Boffi, F. Brezzi, and L. Gastaldi. On the convergence of eigenvalues for mixed formulations. *Ann. Scuola Norm. Sup. Pisa Cl. Sci.* (4), 25:131–154, 1997.

[4] D. Boffi, M. Conforti, and L. Gastaldi. Modified edge finite elements for photonic crystals. *Numerische Mathematik*, 105(2):249–266, 2006.

[5] D. Boffi, L. Demkowicz, and M. Costabel. Discrete compactness for p and hp 2d edge finite elements. *Math. Models Methods Appl. Sci.*, 13:1673–1687, 2003.

[6] D. Boffi and L. Gastaldi. Interpolation estimates for edge finite elements and application to band gap computation. *Appl. Numer. Math.*, 56:1283–1292, 2006.

[7] A. Bonito and J.-L. Guermond. Approximation of the eigenvalue problem for the time harmonic Maxwell system by continuous Lagrange finite elements, 2009. Report IAMCS 2009–121, Texas A&M.

[8] S. C. Brenner and L. R. Scott. *The mathematical theory of finite element methods*, volume 15 of *Texts in Applied Mathematics*. Springer, New York, 1994.

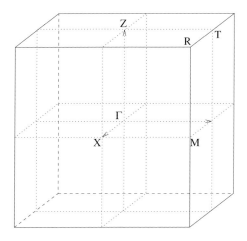

Figure 2.2: The Brillouin zone and the points of high symmetry.

[9] A. Bulovyatov. *Parallel multigrid methods for the band structure computation of 3D photonic crystals with higher order finite elements.* Ph.D. thesis, Karlsruhe Institute of Technology, 2010.

[10] S. Caorsi, P. Fernandes, and M. Raffetto. On the convergence of Galerkin finite element approximations of electromagnetic eigenproblems. *SIAM J. Numer. Anal.*, 38:580–607, 2000.

[11] P. Ciarlet Jr. and G. Hechme. Computing electromagnetic eigenmodes with continuous Galerkin approximations. *Comput. Methods Appl. Mech. Engrg.*, 198:358–365, 2008.

[12] M. Costabel and M. Dauge. Maxwell and Lamé eigenvalues on polyhedra. *Math. Methods Appl. Sci.*, 22:243–258, 1999.

[13] M. Costabel and M. Dauge. Weighted regularization of Maxwell equations in polyhedral domains. A rehabilitation of nodal finite elements. *Numer. Math.*, 93:239–277, 2002.

[14] M. Costabel, M. Dauge, and S. Nicaise. Singularities of Maxwell interface problems. *Math. Model. Numer. Anal.*, 33:627–649, 1999.

[15] L. Demkowicz, P. Monk, C. Schwab, and L. Vardapetyan. Maxwell eigenvalues and discrete compactness in two dimensions. *Comput. Math. Appl.*, 40:589–605, 2000.

[16] D. C. Dobson, Jayadeep Gopalakrishnan, and J. E. Pasciak. An efficient method for band structure calculations in 3D photonic crystals. *Journal of Computational Physics*, 161(2):668–679, 2000.

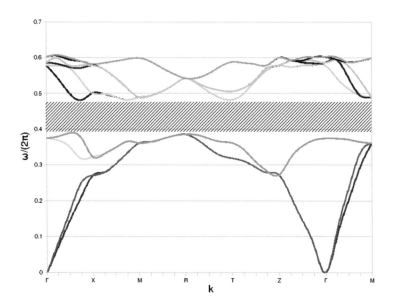

Figure 2.3: Band structure illustration along a curve in the Brillouin zone along the points Γ, X, M, R, T, and Z (cf. Fig. 2.2). A band gap is observed between the bands 4 and 5.

[17] A. Ern and J.-L. Guermond. *Theory and practice of finite elements*, volume 159 of *Applied Mathematical Sciences*. Springer, New York, 2004.

[18] R. Hiptmair. Multigrid method for Maxwell's equations. *SIAM Journal on Numerical Analysis*, 36(1):204–225, 1998.

[19] R. Hiptmair. Finite elements in computational electromagnetism. *Acta Numerica*, 11:237–339, 2002.

[20] R. Hiptmair and K. Neymeyr. Multilevel method for mixed eigenproblems. *SIAM Journal on Scientific Computing*, 23(6):2141–2164, 2002.

[21] V. Hoang, M. Plum, and C. Wieners. A computer-assisted proof for photonic band gaps. *Zeitschrift für Angewandte Mathematik und Physik*, 60:1–18, 2009.

[22] F. Kikuchi. On a discrete compactness property for the Nédélec finite elements. *J. Fac. Sci. Univ. of Tokio Sec. IA*, 36:479–490, 1989.

[23] F. Kikuchi. Mixed Formulations for Finite Element Analysis of Magnetostatic and Electrostatic Problems. *Japan J. Appl. Math.*, 6:209–221, 1989.

[24] A. V. Knyazev. Toward the optimal preconditioned eigensolver: locally optimal block preconditioned conjugate gradient method. *SIAM Journal on Scientific Computing*, 23(2):517–541, 2001.

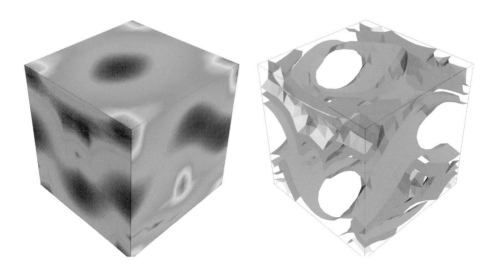

Figure 2.4: Illustration of the eigenfunction $\boldsymbol{u}_{h,\boldsymbol{k}}^5$. On the left, a surface plot of the amplitude of $|\boldsymbol{u}_{h,\boldsymbol{k}}^5|$ and on the right isosurfaces are displayed.

[25] A. V. Knyazev and K. Neymeyr. A geometric theory for preconditioned inverse iteration. III: A short and sharp convergence estimate for generalized eigenvalue problems. *Linear Algebra and its Applications*, 358(1):95–114, 2003.

[26] P. Monk. *Finite Element Methods for Maxwell's Equations*. Clarendon Press, Oxford, 2003.

[27] J. Schöberl. Commuting quasi interpolation operators for mixed finite elements, 2002. Report ISC-01-10-MATH, Texas A&M.

[28] C. Wieners. Distributed point objects. A new concept for parallel finite elements. In R. Kornhuber, R. Hoppe, J. Priaux, O. Pironneau, O. Widlund, and J. Xu, editors, *Domain Decomposition Methods in Science and Engineering*, volume 40 of *Lecture Notes in Computational Science and Engineering*, pages 175–183. Springer, 2004.

[29] C. Wieners. A geometric data structure for parallel finite elements and the application to multigrid methods with block smoothing. *Computing and Visualization in Science*, 13:161–175, 2010.

[30] S. Zaglmayr. *High order finite element methods for electromagnetic field computation*. PhD thesis, Johannes Kepler Universität Linz, 2006.

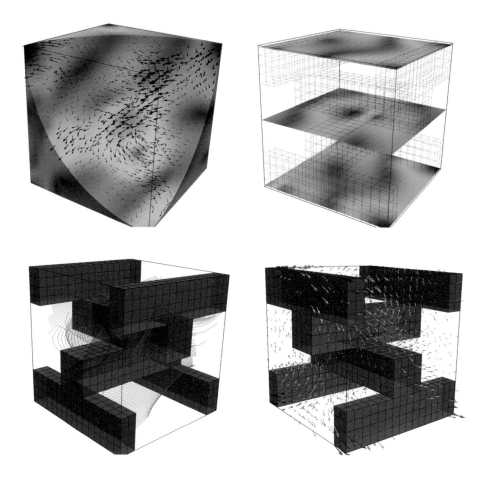

Figure 2.5: Illustration of the eigenfunction $\boldsymbol{u}_{h,\boldsymbol{k}}^6$ for $\boldsymbol{k} = (3, 1, -2)$. The cutting planes (top row) show the amplitude of the magnetic energy distribution $|\boldsymbol{u}_{h,\boldsymbol{k}}^6|$, the streamlines and the arrow-plot (bottom row) represent the vector field $\boldsymbol{u}_{h,\boldsymbol{k}}^6$.

Chapter 3

On the spectra of periodic differential operators

by Michael Plum

The main mathematical tool for treating spectral problems for differential operators with periodic coefficients is the so-called Floquet–Bloch theory. It relates the spectrum of a selfadjoint operator realising a periodic spectral problem on the whole of $\mathbb{R}^n$ to a family of eigenvalue problems on the periodicity cell. Here, the eigenfunctions ("Bloch waves") are subject to semi-periodic boundary conditions depending on an additional parameter which varies over the so-called Brillouin zone. The result is the "band-gap" structure of the spectrum of the whole-space operator. Floquet–Bloch theory applies, in particular, to periodic Schrödinger equations and — what is most important within the scope of this book — to periodic Maxwell eigenvalue problems, i. e. to photonic crystals.

This theory is well-known to all experts, but it is not easy to find a self-contained exposition which moreover uses elementary arguments giving an easy access also for (doctoral) students. Since this book is mainly aiming at educating doctoral students, such a self-contained description is given here. On one hand, it is rather general because periodic differential operators of arbitrary even order are considered (which actually does not complicate the arguments), but on the other hand the coefficients are assumed to be "smooth", in order to satisfy our requirement of easy access. A more general exposition for non-smooth coefficients has been published in [2]. Important contributions to Floquet–Bloch theory have been made e. g. in [9, 10, 11].

Floquet–Bloch theory does not give an answer to the question if there are really gaps in the spectrum or if the bands actually overlap. An asymptotic answer (for sufficiently "high contrast" in the coefficients) has been given in [5]. In a more concrete case, existence of a gap has been proved by computer-assisted means in [8].

3.1   The spectrum of selfadjoint operators

Here, we briefly summarise some basic results about the spectrum of selfadjoint linear operators

Definition. Let $(H, \langle \cdot, \cdot \rangle)$ denote a complex Hilbert space, $D(A) \subset H$ a dense subspace, and $A : D(A) \to H$ a linear operator.

a) The *adjoint* $A^* : D(A^*) \to H$ of A is defined by
   $D(A^*) := \{u \in H : \exists_{u^* \in H} \forall_{v \in D(A)} \langle u, Av \rangle = \langle u^*, v \rangle\}$ and $A^*u := u^*$ for $u \in D(A^*)$; note that, for $u \in D(A^*)$, u^* is unique.

b) A is *selfadjoint* iff $A = A^*$ (i.e. $D(A) = D(A^*)$, $Au = A^*u$ for $u \in D(A)$).

Definition. Let $A : D(A) \to H$ be a selfadjoint linear operator.

a) The *resolvent set* $\rho(A) \subset \mathbb{C}$ of A is defined as $\rho(A) := \{\lambda \in \mathbb{C} : A - \lambda I : D(A) \to H$ is bijective$\}$.

b) The *spectrum* of A is the set $\sigma(A) := \mathbb{C} \setminus \rho(A)$.

c) The *point spectrum* $\sigma_p(A)$ of A is the set of all eigenvalues of A.

d) The *continuous spectrum* $\sigma_c(A)$ of A is the set $\sigma_c(A) := \{\lambda \in \mathbb{C} : A - \lambda I$ is one-to-one but not onto$\}$.

Basic results.

i) For $\lambda \in \rho(A)$, $(A - \lambda I)^{-1}$ is bounded.

Proof. A is selfadjoint and hence closed, which implies that $(A - \lambda I)^{-1}$ is closed. Moreover, $(A - \lambda I)^{-1}$ is defined on H, whence its boundedness follows from the Closed Graph Theorem. □

ii) For all $\lambda \in \mathbb{C}$, $\mathrm{kernel}\,(A - \lambda I) = \mathrm{range}\,(A - \lambda I)^{\perp}$.

Proof. If $u \in \mathrm{kernel}\,(A - \lambda I) \setminus \{0\}$, we have $\lambda \in \mathbb{R}$ since A is symmetric, and $\langle u, (A - \lambda I)v \rangle = \langle (A - \lambda I)u, v \rangle = 0$ for all $v \in D(A)$, i.e. $u \in \mathrm{range}\,(A - \lambda I)^{\perp}$. If vice versa $u \in \mathrm{range}\,(A - \lambda I)^{\perp} \setminus \{0\}$, we obtain $\langle u, Av \rangle = \langle \bar{\lambda} u, v \rangle$ for all $v \in D(A)$, whence the selfadjointness gives $u \in D(A)$, $Au = \bar{\lambda} u$, and $\lambda \in \mathbb{R}$ since A is symmetric, i.e. $u \in \mathrm{kernel}\,(A - \lambda I)$. □

iii) For $\lambda \in \sigma_c(A)$, $\mathrm{range}\,(A - \lambda I)$ is dense in H (but not equal to H), and $(A - \lambda I)^{-1}$ is unbounded.

Proof. Since $A - \lambda I$ is one-to-one, $\mathrm{range}\,(A - \lambda I)$ is dense in H by ii). If $(A - \lambda I)^{-1}$ were bounded, then (since $(A - \lambda I)^{-1}$ is closed) $\mathrm{range}\,(A - \lambda I)$ would be closed, and hence equal to H, which contradicts the definition of $\sigma_c(A)$. □

iv) $\sigma(A) = \sigma_c(A) \dot{\cup} \sigma_p(A)$.

Proof. For each $\lambda \in \mathbb{C}$, $A - \lambda I$ is either bijective (i.e. $\lambda \in \rho(A)$), or one-to-one but not onto (i.e. $\lambda \in \sigma_c(A)$), or not one-to-one (i.e. $\lambda \in \sigma_p(A)$). This gives the result. $\qquad\square$

v) $\sigma(A) \subset \mathbb{R}$

Proof. For $\lambda = \mu + i\nu \in \mathbb{C}$, $\nu \neq 0$, we calculate $\|(A - \lambda I)u\|^2 = \|(A - \mu I)u\|^2 + \nu^2 \|u\|^2 \geq \nu^2 \|u\|^2$ for all $u \in D(A)$, which shows that $A - \lambda I$ is one-to-one, and $(A - \lambda I)^{-1}$ is bounded. So $\lambda \notin \sigma_p(A)$, and iii) shows that $\lambda \notin \sigma_c(A)$. Thus, $\lambda \in \rho(A)$ by iv). $\qquad\square$

3.2 Periodic differential operators

Let $L(x, D) = \sum_{|\alpha| \leq m} c_\alpha(x) D^\alpha$ denote an m-th order linear differential operator on $\mathbb{R}^N$ with complex-valued and *periodic coefficients*, i.e. there exist linearly independent vectors $a_1, \ldots, a_N \in \mathbb{R}^N$ such that

$$c_\alpha(x + a_j) = c_\alpha(x) \qquad (x \in \mathbb{R}^N, |\alpha| \leq m, j = 1, \ldots, N). \tag{3.1}$$

We assume that $L(\cdot, D)$ is uniformly strongly elliptic, i.e. m is even, and there exists $\delta > 0$ such that

$$(-1)^{\frac{m}{2}} \operatorname{Re}\left\{ \sum_{|\alpha|=m} c_\alpha(x) \xi^\alpha \right\} \geq \delta |\xi|^m \text{ for all } x, \xi \in \mathbb{R}^N.$$

(Note that the following assumption, together with (3.1), implies boundedness of the coefficients.) We suppose that

$$c_\alpha \in C^{|\alpha|}(\mathbb{R}^N) \text{ for } \alpha \neq (0, \ldots, 0), \ c_{(0, \ldots, 0)} \in L^\infty(\mathbb{R}^N), \tag{3.2}$$

and that $L(\cdot, D)$ is *formally symmetric*, i.e. it coincides with its formal adjoint $L^*(\cdot, D)$ given by $L^*(\cdot, D)u = \sum_{|\alpha| \leq m} (-1)^{|\alpha|} D^\alpha(\bar{c}_\alpha u)$. This implies

$$\int_{\mathbb{R}^N} [L(\cdot, D)u] \, \bar{v} dx = \int_{\mathbb{R}^N} u \overline{L(\cdot, D)v} dx \tag{3.3}$$

for all $u, v \in C_0^\infty(\mathbb{R}^N)$, and hence, by denseness, for all $u, v \in H^m(\mathbb{R}^N)$. See also Chapter 1.2 for a description of the Lebesgue and Sobolev function spaces used here and in the following.

Furthermore, let $w \in L^\infty(\mathbb{R}^N)$ denote some real-valued weight function which is bounded from below by a positive constant, and which is also periodic in the sense (3.1). Thus, the complex Hilbert space $L^2(\mathbb{R}^N)$ can be endowed with the weighted scalar product

$$\langle u, v \rangle := \int_{\mathbb{R}^N} w(x) u(x) \overline{v(x)} \, dx \qquad (u, v \in L^2(\mathbb{R}^N))$$

which is equivalent to the canonical (unweighted) one.

Now define an operator A in $L^2(\mathbb{R}^N)$ by

$$D(A) := H^m(\mathbb{R}^N), \quad Au := \frac{1}{w} L(\cdot, D)u. \tag{3.4}$$

Lemma 3.2.1. *A is selfadjoint.*

Proof. By (3.3), (3.4), A is symmetric on $H^m(\mathbb{R}^N)$ with respect to $\langle \cdot, \cdot \rangle$, which implies that $H^m(\mathbb{R}^N) \subset D(A^*)$ and $Au = A^*u$ for $u \in H^m(\mathbb{R}^N)$. To prove the reverse inclusion $D(A^*) \subset H^m(\mathbb{R}^N)$, let $u \in D(A^*)$, i.e. $u \in L^2(\mathbb{R}^N)$ and there exists some $u^* \in L^2(\mathbb{R}^N)$ such that

$$\langle u, Av \rangle = \langle u^*, v \rangle \quad \text{for all} \ \ v \in H^m(\mathbb{R}^N). \tag{3.5}$$

Now let Ω denote some fundamental domain of periodicity (see also Section 3), and $\Omega_0 \supset \bar{\Omega}$ some additional bounded domain. $\mathbb{R}^N$ is the union (with measure zero overlap) of countably many translation copies $\bar{\Omega} + z_n$ ($n \in \mathbb{N}$) of $\bar{\Omega}$. (3.5) and (3.4) imply, for each $n \in \mathbb{N}$,

$$\int_{\Omega_0 + z_n} u \overline{L(\cdot, D)v} dx = \int_{\Omega_0 + z_n} w u^* \bar{v} dx \quad \text{for all} \ \ v \in C_0^\infty(\Omega_0 + z_n),$$

whence the transformation $x \to x - z_n$ and the periodicity of the coefficients give, for each $n \in \mathbb{N}$,

$$\int_{\Omega_0} u(\cdot + z_n) \overline{L(\cdot, D)v} dx = \int_{\Omega_0} w u^*(\cdot + z_n) \bar{v} dx \quad \text{for all} \ \ v \in C_0^\infty(\Omega_0).$$

Therefore, by [1, Thm. 6.3],

$$u(\cdot + z_n)\big|_\Omega \in H^m(\Omega),$$
$$\|u(\cdot + z_n)\|_{H^m(\Omega)} \leq \gamma \left(\|u^*(\cdot + z_n)\|_{L^2(\Omega_0)} + \|u(\cdot + z_n)\|_{L^2(\Omega_0)} \right), \tag{3.6}$$

where γ depends only on $L(\cdot, D), \Omega, \Omega_0$, and w. Squaring this inequality and transforming $x \to x + z_n$ in the integrals in the norms we obtain

$$\|u\|_{H^m(\Omega + z_n)}^2 \leq 2\gamma^2 \left(\|u^*\|_{L^2(\Omega_0 + z_n)}^2 + \|u\|_{L^2(\Omega_0 + z_n)}^2 \right).$$

Here, the right-hand side is summable over $n \in \mathbb{N}$ since Ω_0 intersects only with finitely many of the (disjoint) translation copies of Ω, and $u, u^* \in L^2(\mathbb{R}^N)$. Hence also the left-hand side is summable, implying $u \in H^m(\mathbb{R}^N)$. $\qquad\square$

Remarks.

a) We can alternatively assume that $L(\cdot, D)$ is given in divergence form $\sum_{|\alpha|,|\beta|\leq m/2} (-1)^{|\alpha|} D^\alpha (c_{\alpha\beta} D^\beta)$. Then the formal symmetry simply reads $c_{\alpha\beta} = \overline{c_{\beta\alpha}}$. Instead of (3.2), we now require

$$c_{\alpha\beta} \in C^{\frac{m}{2}}(\mathbb{R}^N) \tag{3.7}$$

which is weaker than (3.2) for the "higher order" coefficients, but stronger for the "lower order" ones. Under these assumptions (together with uniform strong ellipticity), we can use [6, Thm. 16.1] (instead of [1]) to obtain the crucial statement (3.6) in the proof of Lemma 3.2.1.

b) Both alternative smoothness assumptions (3.2) and (3.7) may be regarded as unpleasant in view of some applications. In [2], a more general (but technically more involved) theory is presented where $L(\cdot, D)$ is given in divergence form, with the coefficients $c_{\alpha\beta}$ assumed to be merely in $L^\infty(\mathbb{R}^N)$ (and periodic). Here, the underlying Hilbert space is chosen to be the dual space $H^{-m/2}(\mathbb{R}^N)$.

c) For a general description of the construction of selfadjoint operators from formally selfadjoint differential expressions, see e. g. the books [3, 7, 11].

Our aim is now to prove a "band-gap" structure of the spectrum of A, i. e. to prove that the spectrum of A is a union of compact intervals. The proof will furthermore give access to computation of these spectral bands via eigenvalue problems on bounded domains.

3.3 Fundamental domain of periodicity and the Brillouin zone

Let Ω be a fundamental domain of periodicity associated with (3.1). For example, Ω can be chosen to be the $N-$dimensional parallelogram which has the origin in $\mathbb{R}^N$ as one corner, and the vectors a_j forming the sides which meet at that corner. But there is a lot of freedom in choosing Ω. (If e. g. $N = 2$, $a_1 = (1,0)^T$, $a_2 = (0,1)^T$, then $\Omega = (0,1)^2$ can be chosen, but as well the parallelogram spanned by $\widetilde{a}_1 = (1,0)^T$, $\widetilde{a}_2 = (1,1)^T$.)

Next, let $b_1, \ldots, b_N \in \mathbb{R}^N$ denote the columns of $2\pi(M^T)^{-1}$, with M denoting the matrix with columns $a_1, \ldots, a_N$. Hence

$$b_l \cdot a_j = 2\pi\delta_{lj} \qquad (l, j = 1, \ldots, N). \tag{3.8}$$

$b_1, \ldots, b_N$ generate a new periodicity lattice in $\mathbb{R}^N$, the so-called *reciprocal lattice*. As before, we can choose a fundamental domain of periodicity. A particular choice is the set of all points in $\mathbb{R}^N$ which are closer to 0 than to any other point in the reciprocal lattice. This set is called *Brillouin zone B*.

3.4   Bloch waves, Floquet transformation

For fixed $k \in \overline{B}$, we consider the eigenvalue problem

$$L(\cdot, D)\psi = \lambda w \psi \quad \text{on } \Omega, \tag{3.9}$$

$$\psi(x + a_j) = e^{ik \cdot a_j} \psi(x) \quad (j = 1, \ldots, N). \tag{3.10}$$

In writing the boundary condition in the form (3.10), we understand ψ extended to the whole of $\mathbb{R}^N$. In fact, (3.10) forms boundary conditions on $\partial\Omega$, so-called *semi-periodic* boundary conditions.

Since $L(\cdot, D)$ is formally symmetric and the coefficients of L, as well as the weight function w, are periodic with domain of periodicity Ω, we conclude that (3.9), (3.10) is a *symmetric* eigenvalue problem in $L^2(\Omega; w)$. Since Ω is bounded, compactness arguments can be used to prove that (3.9), (3.10) has a $\langle \cdot, \cdot \rangle$-orthonormal and complete system $(\psi_s(\cdot, k))_{s \in \mathbb{N}}$ of eigenfunctions in $H_{loc}^m(\mathbb{R}^N)$, with corresponding eigenvalues satisfying

$$\lambda_1(k) \le \lambda_2(k) \le \cdots \le \lambda_s(k) \to \infty \text{ as } s \to \infty. \tag{3.11}$$

The eigenfunctions $\psi_s(\cdot, k)$ are called *Bloch waves*. They can be chosen such that they depend on k in a measurable way (see [11, XIII.16, Theorem XIII.98]).

Now define

$$\phi_s(x, k) := e^{-ik \cdot x} \psi_s(x, k). \tag{3.12}$$

Then,

$$D^\alpha \psi_s(\cdot, k) = e^{ik \cdot x}(D + ik)^\alpha \phi_s(\cdot, k)$$

and thus

$$L(\cdot, D)\psi_s(\cdot, k) = e^{ik \cdot x} L(\cdot, D + ik)\phi_s(\cdot, k). \tag{3.13}$$

Furthermore, using (3.10) and (3.12),

$$\phi_s(x + a_j, k) = e^{-ik \cdot (x + a_j)} \psi_s(x + a_j, k) = \phi_s(x, k),$$

which together with (3.13) shows that $(\phi_s(\cdot, k))_{s \in \mathbb{N}}$ is an orthonormal and complete system of eigenfunctions of the *periodic* eigenvalue problem

$$L(\cdot, D + ik)\phi = \lambda w \phi \quad \text{on } \Omega, \tag{3.14}$$

$$\phi(x + a_j) = \phi(x) \quad (j = 1, \ldots, N),$$

with the same eigenvalue sequence $(\lambda_s(k))_{s \in \mathbb{N}}$ as before. We shall see that the spectrum of the operator A can be constructed from the eigenvalue sequences $(\lambda_s(k))_{s \in \mathbb{N}}$ by varying k over the Brillouin zone B.

An important step towards this aim is the *Floquet transformation*

$$(Uf)(x,k) := \frac{1}{\sqrt{|B|}} \sum_{n\in\mathbb{Z}^N} f(x-Mn)e^{ik^T Mn} \qquad (x\in\Omega, k\in B), \qquad (3.15)$$

recalling that M denotes the matrix with columns $a_1,\ldots,a_N$.

Lemma 3.4.1. $U : L^2(\mathbb{R}^N) \to L^2(\Omega\times B)$ *is an isometric isomorphism, with inverse*

$$(U^{-1}g)(x-Mn) = \frac{1}{\sqrt{|B|}} \int_B g(x,k)e^{-ik^T Mn}\,dk \qquad (x\in\Omega, n\in\mathbb{Z}^N). \quad (3.16)$$

If $g(\cdot,k)$ is extended to the whole of $\mathbb{R}^N$ by the semi-periodicity condition (3.10), we have

$$U^{-1}g = \frac{1}{\sqrt{|B|}} \int_B g(\cdot,k)\,dk. \qquad (3.17)$$

Proof. For $f \in L^2(\mathbb{R}^N)$,

$$\int_{\mathbb{R}^N} w|f(x)|^2\,dx = \sum_{n\in\mathbb{Z}^N} \int_\Omega w|f(x-Mn)|^2\,dx. \qquad (3.18)$$

Here, we can exchange summation and integration by Beppo Levi's Theorem. Therefore,

$$\sum_{n\in\mathbb{Z}^N} |f(x-Mn)|^2 < \infty \quad \text{for a. e. } x\in\Omega.$$

Thus, $(Uf)(x,k)$ is well defined by (3.15) (as a Fourier series with variable $M^T k$) for a. e. $x\in\Omega$, and Parseval's equality gives, for these x,

$$\int_B |(Uf)(x,k)|^2\,dk = \sum_{n\in\mathbb{Z}^N} |f(x-Mn)|^2.$$

By (3.18), this expression is in $L^2(\Omega)$, and

$$\|Uf\|_{L^2(\Omega\times B)} = \|f\|_{L^2(\mathbb{R}^N)}.$$

We are left to show that U is onto, and that U^{-1} is given by (3.16) or (3.17). Let $g \in L^2(\Omega\times B)$, and define

$$f(x-Mn) := \frac{1}{\sqrt{|B|}} \int_B g(x,k)e^{-ik^T Mn}\,dk \qquad (x\in\Omega, n\in\mathbb{Z}^N). \qquad (3.19)$$

For fixed $x \in \Omega$, Plancherel's Theorem gives

$$\sum_{n \in \mathbb{Z}^N} |f(x - Mn)|^2 = \int_B |g(x,k)|^2 \, dk \,,$$

whence, by integration over Ω,

$$\int_{\Omega \times B} w|g(x,k)|^2 \, dx dk = \int_\Omega \sum_{n \in \mathbb{Z}^N} w|f(x - Mn)|^2 \, dx$$

$$= \sum_{n \in \mathbb{Z}^N} \int_\Omega w|f(x - Mn)|^2 \, dx = \int_{\mathbb{R}^N} w|f(x)|^2 \, dx \,,$$

i.e. $f \in L^2(\mathbb{R}^N)$. Now (3.15) gives, for a.e. $x \in \Omega$,

$$f(x - Mn) = \frac{1}{\sqrt{|B|}} \int_B (Uf)(x,k) e^{-ik^T Mn} \, dk \qquad (n \in \mathbb{Z}^N) \,,$$

whence (3.19) implies $Uf = g$ and (3.16). Now (3.17) follows from (3.16) using $g(x + Mn, k) = e^{ik^T Mn} g(x, k)$. $\qquad\square$

3.5   Completeness of the Bloch waves

Using the Floquet transformation U, we are now able to prove a completeness property of the Bloch waves $\psi_s(\cdot, k)$ in $L^2(\mathbb{R}^N)$ when we vary k over the Brillouin zone B.

Theorem 3.5.1. *For each $f \in L^2(\mathbb{R}^N)$ and $l \in \mathbb{N}$, define*

$$f_l(x) := \frac{1}{\sqrt{|B|}} \sum_{s=1}^l \int_B \langle (Uf)(\cdot, k), \psi_s(\cdot, k) \rangle_{L^2(\Omega;w)} \, \psi_s(x, k) \, dk \quad (x \in \mathbb{R}^N). \quad (3.20)$$

Then, $f_l \to f$ in $L^2(\mathbb{R}^N)$ as $l \to \infty$.

Proof. Since $Uf \in L^2(\Omega \times B)$, we have $(Uf)(\cdot, k) \in L^2(\Omega)$ for a.e. $k \in B$ by Fubini's Theorem. Since $(\psi_s(\cdot, k))_{s \in \mathbb{N}}$ is orthonormal and complete in $L^2(\Omega; w)$ for each $k \in B$, we obtain

$$\lim_{l \to \infty} \|(Uf)(\cdot, k) - g_l(\cdot, k)\|_{L^2(\Omega;w)} = 0 \quad \text{for a.e. } k \in B$$

where

$$g_l(x, k) := \sum_{s=1}^l \langle (Uf)(\cdot, k), \psi_s(\cdot, k) \rangle_{L^2(\Omega;w)} \, \psi_s(x, k) \,. \quad (3.21)$$

Thus, for $\chi_l(k) := \|(Uf)(\cdot, k) - g_l(\cdot, k)\|^2_{L^2(\Omega;w)}$, we get

$$\chi_l(k) \to 0 \text{ as } l \to \infty \quad \text{for a. e. } k \in B\,,$$

and moreover, by Bessel's inequality,

$$\chi_l(k) \le \|(Uf)(\cdot, k)\|^2_{L^2(\Omega;w)} \quad \text{for all } l \in \mathbb{N} \text{ and a. e. } k \in B\,,$$

and $\|(Uf)(\cdot, k)\|^2_{L^2(\Omega;w)}$ is in $L^1(B)$ as a function of k, by Lemma 3.4.1. Altogether, Lebesgue's Dominated Convergence Theorem implies

$$\int_B \chi_l(k)\, dk \to 0 \text{ as } l \to \infty\,,$$

i. e.,

$$\|Uf - g_l\|_{L^2(\Omega \times B)} \to 0 \text{ as } l \to \infty\,. \tag{3.22}$$

Using (3.20), (3.21), and (3.17), we find that $f_l = U^{-1}g_l$, whence (3.22) gives

$$\|U(f - f_l)\|_{L^2(\Omega \times B)} \to 0 \text{ as } l \to \infty\,,$$

and the assertion follows since $U : L^2(\mathbb{R}^N) \to L^2(\Omega \times B)$ is isometric by Lemma 3.4.1. □

3.6 The spectrum of A

In this section, we will prove the main result stating that

$$\sigma(A) = \bigcup_{s \in \mathbb{N}} I_s\,, \tag{3.23}$$

where

$$I_s := \{\lambda_s(k) : k \in \overline{B}\} \quad (s \in \mathbb{N})\,. \tag{3.24}$$

For each $s \in \mathbb{N}$, λ_s is a continuous function of $k \in \overline{B}$, which follows by standard arguments from the fact that the coefficients in the eigenvalue problem (3.14) depend continuously on k. Thus, since $\overline{B}$ is compact and connected,

$$I_s \text{ is a compact real interval, for each } s \in \mathbb{N}\,. \tag{3.25}$$

Moreover, Poincaré's min-max principle for eigenvalues implies that

$$\mu_s \le \lambda_s(k) \quad \text{for all } s \in \mathbb{N}, k \in \overline{B}\,,$$

with $(\mu_s)_{s\in\mathbb{N}}$ denoting the sequence of eigenvalues of problem (3.9) with *Neumann* ("free") boundary conditions. Since $\mu_s \to \infty$ as $s \to \infty$, we obtain

$$\min I_s \to \infty \text{ as } s \to \infty, \tag{3.26}$$

which together with (3.25) implies that

$$\bigcup_{s\in\mathbb{N}} I_s \text{ is closed.} \tag{3.27}$$

The first part of the statement (3.23) is

Theorem 3.6.1. $\sigma(A) \supset \bigcup_{s\in\mathbb{N}} I_s$.

Proof. Let $\lambda \in \bigcup_{s\in\mathbb{N}} I_s$, i.e. $\lambda = \lambda_s(k)$ for some $s \in \mathbb{N}$ and some $k \in \overline{B}$, and

$$L(\cdot, D)\psi_s(\cdot, k) = \lambda w\psi_s(\cdot, k). \tag{3.28}$$

We regard $\psi_s(\cdot, k)$ as extended to the whole of $\mathbb{R}^N$ by the boundary condition (3.10), whence, due to the periodicity of the coefficients of $L(\cdot, D)$, (3.28) holds for all $x \in \mathbb{R}^N$.

We choose a function $\eta \in C_0^\infty(\mathbb{R}^N)$ such that

$$\eta(x) = 1 \quad \text{for } |x| \leq 1, \quad \eta(x) = 0 \quad \text{for } |x| \geq 2,$$

and define, for each $l \in \mathbb{N}$,

$$u_l(x) := \eta\left(\frac{|x|}{l}\right)\psi_s(x, k).$$

Then,

$$(L(\cdot, D) - \lambda w)u_l = \sum_{|\alpha|\leq m} c_\alpha D^\alpha\left[\eta\left(\frac{|\cdot|}{l}\right)\psi_s(\cdot, k)\right] - \lambda w\eta\left(\frac{|\cdot|}{l}\right)\psi_s(\cdot, k) \tag{3.29}$$

$$= \eta\left(\frac{|\cdot|}{l}\right)(L(\cdot, D) - \lambda w)\psi_s(\cdot, k) + R,$$

where R is a sum of products of bounded functions, derivatives (of order ≥ 1) of $\eta\left(\frac{|\cdot|}{l}\right)$, and derivatives (of order $\leq m - 1$) of $\psi_s(\cdot, k)$. Thus (note that $\psi_s(\cdot, k) \in H_{\text{loc}}^m(\mathbb{R}^N)$),

$$\|R\|_{L^2(\mathbb{R}^N)} \leq \frac{c}{l}\|\psi_s(\cdot, k)\|_{H^{m-1}(K_{2l})}, \tag{3.30}$$

with K_{2l} denoting the ball in $\mathbb{R}^N$ with radius $2l$, centered at 0. Moreover, the semi-periodic structure of $\psi_s(\cdot, k)$ implies

$$\|\psi_s(\cdot, k)\|_{H^{m-1}(K_{2l})} \leq c\sqrt{\text{vol}(K_{2l})}.$$

Together with (3.28), (3.29), (3.30), this gives

$$\|(L(\cdot, D) - \lambda w)u_l\|_{L^2(\mathbb{R}^N)} \leq \frac{c}{l}\sqrt{\mathrm{vol}(K_{2l})}\,.$$

Furthermore, again by the semiperiodic structure of $\psi_s(\cdot, k)$,

$$\|u_l\|_{L^2(\mathbb{R}^N)} \geq c\|\psi_s(\cdot, k)\|_{L^2(K_l)} \geq c\sqrt{\mathrm{vol}(K_l)}$$

with $c > 0$. Since $\mathrm{vol}(K_{2l})/\mathrm{vol}(K_l)$ is bounded, we obtain

$$\frac{1}{\|u_l\|_{L^2(\mathbb{R}^N)}}\|(L(\cdot, D) - \lambda w)u_l\|_{L^2(\mathbb{R}^N)} \leq \frac{c}{l}\,.$$

Because moreover $u_l \in H^m(\mathbb{R}^N) = D(A)$, this results in

$$\frac{1}{\|u_l\|_{L^2(\mathbb{R}^N)}}\|(A - \lambda I)u_l\|_{L^2(\mathbb{R}^N)} \to 0 \text{ as } l \to \infty\,.$$

Thus, either λ is an eigenvalue of A, or $(A - \lambda I)^{-1}$ exists but is unbounded. In both cases, $\lambda \in \sigma(A)$ by the result i) in Section 1. $\qquad\square$

Now we turn to the reverse statement

Theorem 3.6.2. $\sigma(A) \subset \bigcup_{s \in \mathbb{N}} I_s$

Proof. Let $\lambda \in \mathbb{R} \setminus \bigcup_{s \in \mathbb{N}} I_s$. We have to prove that $\lambda \in \rho(A)$, i.e., that, for each $f \in L^2(\mathbb{R}^N)$, some $u \in D(A)$ exists satisfying $(A - \lambda I)u = f$. (Then, $A - \lambda I$ is onto, and hence also one-to-one by the basic result ii) in Section 1.) For given $f \in L^2(\mathbb{R}^N)$, we define, for $l \in \mathbb{N}$,

$$f_l(x) := \frac{1}{\sqrt{|B|}} \sum_{s=1}^{l} \int_B \langle (Uf)(\cdot, k), \psi_s(\cdot, k) \rangle_{L^2(\Omega;w)} \psi_s(x, k)\, dk$$

and

$$u_l(x) := \frac{1}{\sqrt{|B|}} \sum_{s=1}^{l} \int_B \frac{1}{\lambda_s(k) - \lambda} \langle (Uf)(\cdot, k), \psi_s(\cdot, k) \rangle_{L^2(\Omega;w)} \psi_s(x, k)\, dk\,. \quad (3.31)$$

Here, we note that, due to (3.27), some $\delta > 0$ exists such that

$$|\lambda_s(k) - \lambda| \geq \delta \quad \text{for all } s \in \mathbb{N},\ k \in B\,. \quad (3.32)$$

In particular, the boundary value problem

$$(L(\cdot, D) - \lambda w)v(\cdot, k) = w(Uf)(\cdot, k) \text{ in } \Omega\,, \quad (3.33)$$

$$v(x + a_j) = e^{ik \cdot a_j} v(x) \qquad (j = 1, \ldots, N)$$

has a unique solution for every $k \in B$. Bloch wave expansion gives

$$\|(Uf)(\cdot,k)\|^2_{L^2(\Omega;w)} = \sum_{s=1}^{\infty} |\langle (Uf)(\cdot,k),\, \psi_s(\cdot,k)\rangle_{L^2(\Omega;w)}|^2$$

$$= \sum_{s=1}^{\infty} |\langle (w^{-1}L(\cdot,D) - \lambda)v(\cdot,k),\, \psi_s(\cdot,k)\rangle_{L^2(\Omega;w)}|^2\,.$$

Since both $v(\cdot,k)$ and $\psi_s(\cdot,k)$ satisfy semi-periodic boundary conditions, $w^{-1}L(\cdot,D) - \lambda$ can be moved to $\psi_s(\cdot,k)$ in the inner product, and hence (3.9) and (3.32) give

$$\|(Uf)(\cdot,k)\|^2_{L^2(\Omega;w)} = \sum_{s=1}^{\infty} |\lambda_s(k) - \lambda|^2 |\langle v(\cdot,k),\, \psi_s(\cdot,k)\rangle_{L^2(\Omega;w)}|^2$$

$$\geq \delta^2 \|v(\cdot,k)\|^2_{L^2(\Omega;w)}\,.$$

By Lemma 3.4.1, this implies $v \in L^2(\Omega \times B)$, and we can define $u := U^{-1}v \in L^2(\mathbb{R}^N)$. Thus, (3.33) gives

$$\langle (Uf)(\cdot,k),\, \psi_s(\cdot,k)\rangle_{L^2(\Omega;w)} = \langle (w^{-1}L(\cdot,D) - \lambda)(Uu)(\cdot,k),\, \psi_s(\cdot,k)\rangle_{L^2(\Omega;w)}$$

$$= \langle (Uu)(\cdot,k),\, (w^{-1}L(\cdot,D) - \lambda)\psi_s(\cdot,k)\rangle_{L^2(\Omega;w)}$$

$$= (\lambda_s(k) - \lambda)\langle Uu(\cdot,k),\, \psi_s(\cdot,k)\rangle_{L^2(\Omega;w)}\,,$$

whence (3.31) implies

$$u_l(x) = \frac{1}{\sqrt{|B|}} \sum_{s=1}^{l} \int_B \langle (Uu)(\cdot,k),\, \psi_s(\cdot,k)\rangle_{L^2(\Omega;w)} \psi_s(x,k)\, dk\,, \qquad (3.34)$$

and Theorem 3.5.1 gives

$$u_l \to u, \quad f_l \to f \quad \text{in } L^2(\mathbb{R}^N)\,. \qquad (3.35)$$

We will now prove that (in the distributional sense)

$$(w^{-1}L(\cdot,D) - \lambda)u_l = f_l \quad \text{for all } l \in \mathbb{N}\,, \qquad (3.36)$$

which implies (compare (3.37) below) that $\langle u_l, (A - \lambda I)v\rangle = \langle f_l, v\rangle$ for all $v \in H^m(\mathbb{R}^N) = D(A)$, whence Lemma 3.2.1 implies $u_l \in D(A)$, and

$$(A - \lambda I)u_l = f_l \quad \text{for all } l \in \mathbb{N}\,.$$

Since A is closed, (3.35) now implies

$$u \in D(A), \text{ and } (A - \lambda I)u = f\,,$$

which is the desired result.

We are left to prove (3.36), i. e. that

$$\langle u_l, (w^{-1}L(\cdot, D) - \lambda)\varphi \rangle_{L^2(\mathbb{R}^N)} = \langle f_l, \varphi \rangle_{L^2(\mathbb{R}^N)} \qquad \text{for all } \varphi \in C_0^\infty(\mathbb{R}^N). \quad (3.37)$$

So let $\varphi \in C_0^\infty(\mathbb{R}^N)$ be fixed, and let $K \subset \mathbb{R}^N$ denote a ball containing $\operatorname{supp}(\varphi)$ in its interior. Both the functions

$$r_s(x, k) := w(x)\frac{1}{\lambda_s(k) - \lambda}\langle (Uf)(\cdot, k), \psi_s(\cdot, k)\rangle_{L^2(\Omega;w)}$$
$$\psi_s(x, k)\overline{(w^{-1}L(x, D) - \lambda)\varphi(x)},$$

$$t_s(x, k) := w(x)\langle (Uf)(\cdot, k), \psi_s(\cdot, k)\rangle_{L^2(\Omega;w)}\psi_s(x, k)\overline{\varphi(x)}$$

are easily seen to be in $L^2(K \times B)$ by Fubini's Theorem, since (3.32), and the fact that $(w^{-1}L(\cdot, D) - \lambda)\varphi \in L^\infty(K)$ and $\varphi \in L^\infty(K)$, imply that both

$$\int_K |r_s(x, k)|^2 \, dx$$

and

$$\int_K |t_s(x, k)|^2 \, dx$$

are bounded by

$$C\|(Uf)(\cdot, k)\|_{L^2(\Omega;w)}^2 \|\psi_s(\cdot, k)\|_{L^2(K)}^2 \, ;$$

the latter factor is bounded as a function of k because K is covered by a finite number of copies of Ω, and the former is in $L^1(B)$ by Lemma 3.4.1.

Since $K \times B$ is bounded, r and t are also in $L^1(K \times B)$. Therefore, Fubini's Theorem implies that the order of integration with respect to x and k may be exchanged for r and t. Thus, by (3.31),

$$\int_K w(x)u_l(x)\overline{(w^{-1}L(x, D) - \lambda)\varphi(x)} \, dx$$

$$= \frac{1}{\sqrt{|B|}}\sum_{s=1}^{l}\int_K \left(\int_B r_s(x, k) \, dk\right) dx$$

$$= \frac{1}{\sqrt{|B|}}\sum_{s=1}^{l}\int_B \frac{1}{\lambda_s(k) - \lambda}\langle (Uf)(\cdot, k), \psi_s(\cdot, k)\rangle_{L^2(\Omega;w)}$$

$$\langle \psi_s(\cdot, k), (w^{-1}L(\cdot, D) - \lambda)\varphi \rangle_{L^2(K;w)} \, dk \, .$$

Since φ has compact support in the interior of K, $w^{-1}L(\cdot, D) - \lambda$ may be moved to $\psi_s(\cdot, k)$, and hence (3.9) gives

$$\int_K w(x)u_l(x)\overline{(w^{-1}L(x,D) - \lambda)\varphi(x)}\,dx$$

$$= \frac{1}{\sqrt{|B|}} \sum_{s=1}^{l} \int_B \langle (Uf)(\cdot, k), \psi_s(\cdot, k)\rangle_{L^2(\Omega;w)} \overline{\langle \psi_s(\cdot, k), \varphi\rangle_{L^2(K;w)}}\,dk$$

$$= \frac{1}{\sqrt{|B|}} \sum_{s=1}^{l} \int_B \left(\int_K t_s(x,k)\,dx \right) dk$$

$$= \int_K w(x) \left[\frac{1}{\sqrt{|B|}} \sum_{s=1}^{l} \int_B \langle (Uf)(\cdot, k), \psi_s(\cdot, k)\rangle_{L^2(\Omega;w)} \psi_s(x, k)\,dk \right] \overline{\varphi(x)}\,dx$$

$$= \int_K w(x)f_l(x)\overline{\varphi(x)}\,dx,$$

i. e. (3.37). $\square$

Theorems 3.6.1 and 3.6.2 give the result (3.23). It is however difficult to decide, whether there are really *gaps* in the union (3.23). Moreover, it is not easy to make statements about the nature of the spectrum $\sigma(A)$, for example to decide if $\sigma(A) = \sigma_c(A)$ (i. e. no eigenvalues occur). The only easy result is

Theorem 3.6.3. $\sigma(A)$ *contains no eigenvalues of finite multiplicity.*

Proof. (See [4]) Let λ be an eigenvalue of A and suppose that $E := \text{kernel}(A - \lambda I)$ is finite dimensional. The periodicity of the coefficients of $L(\cdot, D)$ shows that $f \in E$ implies $f(\cdot + a_1) \in E$. Thus, the mapping

$$V : \left\{ \begin{array}{c} E \to E \\ f \mapsto f(\cdot + a_1) \end{array} \right\}$$

is well defined, and moreover

$$\langle Vf, Vg\rangle_{L^2(\mathbb{R}^N)} = \langle f, g\rangle_{L^2(\mathbb{R}^N)} \text{ for } f, g \in E,$$

i. e. V is unitary. The assumption $\dim E < \infty$ implies that V has at least one eigenvalue $\kappa \in \mathbb{C}$, and $|\kappa| = 1$ since V is unitary. An eigenfunction $f \in E \setminus \{0\}$ of V, associated with κ, satisfies $f(\cdot + a_1) \equiv \kappa f$, and thus $|f(x + a_1)| = |f(x)|$ for all $x \in \mathbb{R}^N$, which contradicts $f \in L^2(\mathbb{R}^N)$. $\square$

Bibliography

[1] S. Agmon. *Lectures on Elliptic Boundary Value Problems*, volume 2 of *Mathematical Studies*. Van Nostrand, Princeton, 1965.

[2] B. M. Brown, V. Hoang, M. Plum, and I. Wood. Floquet–Bloch theory for elliptic problems with discontinuous coefficients. In *Operator Theory: Advances and Applications*, volume 214, pages 1–20. Springer Basel AG, 2011.

[3] N. Dunford and J. T. Schwartz. *Linear Operators I, II*. Interscience, New York and London, 1958, 1963.

[4] M. S. P. Eastham. *The Spectral Theory of Periodic Differential Equations*. Scottish Academic Press, Edinburgh and London, 1973.

[5] A. Figotin and P. Kuchment. Band gap structure of spectra of periodic dielectric and acoustic media. II. Two-dimensional photonic crystals. *SIAM J. Appl. Math.*, 56:1561–1620, 1996.

[6] A. Friedman. *Partial Differential Equations*. Holt, Rinehart and Winston, New York, 1969.

[7] I. M. Glazman. *Direct Methods of Qualitative Spectral Analysis of Singular Differential Operators*. Israel Program for Scientific Translations. Daniel Davey & Co., Inc., Jerusalem, New York, 1965, 1966.

[8] V. Hoang, M. Plum, and C. Wieners. A computer-assisted proof for photonic band gaps. *Zeitschrift für Angewandte Mathematik und Physik*, 60:1–18, 2009.

[9] P. Kuchment. *Floquet theory for partial differential equations*, volume 60 of *Operator Theory, Advances and Applications*. Birkhäuser Verlag, Basel, 1993.

[10] F. Odeh and J. B. Keller. Partial differential equations with periodic coefficients and Bloch waves in crystals. *J. Math. Phys.*, 5:1499–1504, 1964.

[11] M. Reed and B. Simon. *Methods of modern mathematical physics I–IV*. Academic Press (Harcourt Brace Jovanovich, Publishers), New York, 1975–1980.

Chapter 4

An introduction to direct and inverse scattering theory

by Armin Lechleiter

Scattering theory deals with perturbation of waves by obstacles. Our viewpoint on wave propagation is therefore somewhat different compared to the one taken in the previous chapters, particularly since we are here interested in wave propagating through unbounded domains. Such problems occur quite naturally in scientific and engineering applications. The most basic example is scattering of a plane wave from a bounded obstacle placed in a homogeneous space. Possible applications include for instance radar technology. As another example, consider diffraction of light waves from a layer with periodic micro structure. Here, the scattering object is a structure which is bounded in one direction and periodic with respect to the other directions. This example indicates a strong link to photonic crystals and periodic differential operators as studied in the previous chapters, but in our example the crystal is not globally periodic. In contrast to the eigenvalue problems for partial differential equations related to band gaps studied in the previous chapters, scattering theory is more related to source problems for differential equations.

To fix some ideas on scattering theory, let us consider the problem of scattering of waves by bounded scattering objects placed in a homogeneous background medium. In this entire chapter, we concentrate on a penetrable inhomogeneous scattering object. Inside the penetrable inhomogeneity, physical characteristics of the medium are different compared to the background medium outside. Hence, a wave propagating through the locally perturbed medium will differ from a corresponding wave propagating through unperturbed medium. Computation of the difference between these two waves, which is called the scattered wave, is a typical direct scattering problem.

In an inverse scattering problem, one is given (partial) measurements of scattered waves and the task is to determine the perturbation of the medium. Problems of this kind have a wide range of application in, e. g., nondestructive testing, exploration of the earth in geophysics, and noninvasive techniques for medical imaging. Direct and inverse scattering problems are intimately connected with each other in the sense that the inverse problem is dependent on the direct problem. However, their characteristic features are quite different: Typically, direct scattering problems possess a unique solution that depends continuously on the data – such problems are called *well-posed*. Additionally, all direct scattering problems we consider here are linear problems, neglecting that nonlinear scattering is also an important issue. On the other hand, the inverse scattering problems arising as identification problems from the direct problems are intrinsically nonlinear and, even worse, they lack continuity properties: Typically, the operator mapping measurements to the scattering object is discontinuous. Such problems are called *ill-posed* and this property causes problems concerning the stability of numerical algorithms to solve inverse scattering problems.

In this chapter, we will only deal with time-harmonic waves having angular frequency $\omega > 0$. Time dependence of all waves throughout the text is hence $\exp(-i\omega t)$, but this dependence is usually suppressed. We denote by c_0 the wave speed in the background medium and plug in the time dependence into the wave equation $c_0^2 \Delta u - \partial_{tt} u = 0$ to find the Helmholtz equation

$$\Delta u + k^2 u = 0 \tag{4.1}$$

with wave number $k^2 = \omega^2/c_0^2$. The same reduction led to the time-harmonic Maxwell's equations in Chapter 1.1.5. As before, we note that it is the real part of $\exp(i\omega t)u$ which has physical significance (and a sinusoidal time dependence).

Despite the fact that scattering theory for Maxwell's equations in three dimensions is the fundamental model for light propagation in dielectric photonic devices, we prefer for this introduction to stick to simpler model problems and work with a scalar Helmholtz equation most of the time. The scalar problems we consider often arise as simplified models from the full Maxwell system (see Chapter 1.1.5) and share with the latter one many (but not all) of its characteristic properties. For this introduction it seems more appropriate to explain important facts on comparatively simple problems rather than going through the technical difficulties of real-world problems. Interested readers will have no problems to go deeper into the subject, using the monographs [41, 33, 19] as a starting point.

Nevertheless, we assume the reader to be familiar with certain concepts of functional and real analysis as well as Sobolev spaces (see Chapter 1.2.1). We use Fredholm theory a couple of times, integrate by parts, and use trace theorems involving fractional Sobolev spaces on the boundary. A probably more exotic issue that we use without much explanation or proof is the concept of unique continuation. Unique continuation results state that a solution to a homogeneous linear differential equation that vanishes in a certain set needs to vanish even in a larger superset. Whenever we use this concept, we indicate where to find a proof of the specific result employed. Finally, we employ elliptic regularity results several times in the text, mostly we only require interior results from [39]. We also require such regularity theorems in spaces of periodic functions, where they follow, roughly speaking, from the standard regularity results by periodic extension. However, we never carry out this argument in detail. Similarly, we use Rellich's compact embedding lemma in Sobolev spaces of periodic functions with periodic boundary conditions without giving a proof.

This text is an enlarged manuscript of two lectures on scattering theory aimed to be understandable for someone without prior knowledge on this matter. This gives some restriction on space and implies that some classical concepts of time-harmonic scattering theory stay on a somewhat informal level, especially, we do not give proofs of several important results. Hopefully, the text nonetheless provides key ideas and concepts.

4.1 Scattering of Time-harmonic Waves

Scattering problems involve partial differential equations posed on unbounded domains. We are going to see later on that such problems require additional "boundary conditions at infinity" – more precisely, we need to prescribe the asymptotic behaviour of a solution to (4.1) as $|x| \to \infty$ to obtain uniqueness of solution. For this task one employs a so-called radiation condition. Roughly speaking, a radiation condition serves to assert a direction to a time-harmonic wave. Assume that u is a solution to the source problem $\Delta u + k^2 u = f$ in the whole space for a source f with compact support D. From a physical point of view, a wave generated in D should travel away from D in the homogeneous exterior to D. A radiation condition provides a tool to determine whether or not a solution to the Helmholtz equation does so. By the way, a solution to the Helmholtz equation that satisfies a radiation condition it therefore often called an *outgoing wave*.

4.1.1 Radiation Conditions – Waves with Direction

As a motivation for the radiation condition in two and three dimensions we first study the one-dimensional situation – in 1D everything is a little more straightforward. Our arguments will at first stay on a formal level where we do not care for correct function spaces. In one dimension, the homogeneous Helmholtz equation is

$$u'' + k^2 u = 0, \qquad k > 0,$$

and it is well known that two linearly independent solutions to this differential equation are $u_{\pm}(x) := \exp(\pm ikx)$. In the time domain, these two solutions correspond to $e^{i(\pm kx - \omega t)}$, respectively, and we note that they are constant on the characteristics $\{(x,t) : \pm kx - \omega t = C\}$, $C \in \mathbb{R}$. Consider first $\exp(-i\omega t)u_{+}$: As t increases, x needs to increase for that (x,t) remains on the same characteristic.

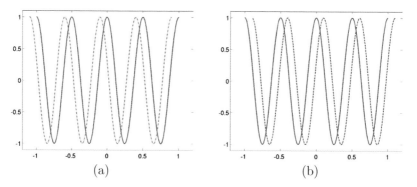

Figure 4.1: In one dimension, a harmonic wave can travel to the left or to the right. For the two plots, we chose $k = \omega = 4\pi$. (a) The solid blue curve shows the real part of $\exp(-i\omega t)u_-$ for $t = 0$ and $x \in (-1, 1)$. The dashed-dotted curve shows the real part of the same function for $t = 0.1$ and $x \in (-1.1, 0.9)$. Obviously, the time-harmonic wave moved to the left. (b) Now we plot the real part of $\exp(-i\omega t)u_+$ for $t = 0$, $x \in (-1, 1)$ (solid curve), and for $t = 0.1$ and $x \in (-0.9, 1.1)$. Here, the curve moves to the right as t increases.

Therefore the wave u_+ moves to the right as time increases. For $\exp(-i\omega t)u_-$, the situation is just the opposite: As t increases, x needs to decrease for that (x, t) remains on the same characteristic and hence u_- moves to the left. This observation is sketched in Figure 4.1. Note that

$$u'_+ - iku_+ = 0 \quad \text{and} \quad u'_- + iku_- = 0 \quad \text{in } \mathbb{R}. \tag{4.2}$$

These two relations can be used to assign a direction to a solution u of $u'' + k^2 u = f$, where f has support in $(-R, R)$ for some $R > 0$. If $u' - iku = 0$ for $x > R$ we say that u travels to the right on the half-axis $\{x \in \mathbb{R} : x > R\}$, and similarly, if $u' + iku = 0$ for $x > R$ we say that u travels to the left in $\{x \in \mathbb{R} : x > R\}$. The first and second option correspond to an outgoing and to an incoming time-harmonic wave, respectively. Of course, the analogous construction works on the left half-axis $\{x \in \mathbb{R} : x > -R\}$.

We finally want to point out how to use the radiation conditions (4.2) to set up a variational solution theory for the outgoing solution $u : \mathbb{R} \to \mathbb{C}$ to $u'' + k^2 u = f$ for, e. g., $f \in L^2(-R, R)$, which is extended by zero to all of $\mathbb{R}$. Actually, we will meet this procedure several times again in the sequel in higher dimensions and for different problems. Since u should have two derivatives more than f, we seek $u \in H^2_{\text{loc}}(\mathbb{R})$. We multiply the differential equation by $\bar{v} \in H^1(-R, R)$ and integrate by parts,

$$\int_{-R}^{R} \left(u'\bar{v}' - k^2 u\bar{v} \right) dx - u'(R)\overline{v(R)} + u'(-R)\overline{v(-R)} = -\int_{-R}^{R} f\bar{v}\, dx. \tag{4.3}$$

Point evaluation of u' and v is well defined since functions in $H^1(a, b)$, $a < b$, are continuous. Since we want u to be outgoing, we prescribe $u'(R) - iku(R) = 0$ and $u'(-R) + iku(-R) = 0$. Thus, (4.3) becomes

$$\mathcal{B}_{1D}(u, v) := \int_{-R}^{R} \left(u'\overline{v'} - k^2 u\overline{v} \right) dx - iku(R)\overline{v(R)} - iku(-R)\overline{v(-R)} \qquad (4.4)$$

$$= -\int_{-R}^{R} f\overline{v}\, dx \qquad (4.5)$$

and we arrive at the following variational formulation: Find $u \in H^1(-R, R)$ such that $\mathcal{B}_{1D}(u, v) = -\int_{-R}^{R} f\overline{v}\, dx$ for all $v \in H^1(-R, R)$. Let us investigate whether this variational problem is solvable. Due to

$$\Re\big(\mathcal{B}_{1D}(u, u)\big) = \int_{-R}^{R} \left(|u'|^2 + |u|^2 \right) dx - (k^2 + 1) \int_{-R}^{R} |u|^2\, dx,$$

the sesquilinear form $\mathcal{B}_{1D}$ is a compact perturbation of a coercive form. Thus, Fredholm's alternative implies that (4.4) has a unique solution in $H^1(-R, R)$ for all $f \in L^2(-R, R)$ if the corresponding homogeneous problem has only the trivial solution. Assume that u is a solution to the homogeneous problem (that is, for $f = 0$). Taking the imaginary part of $\mathcal{B}_{1D}(u, u)$ shows that $k(|u(R)|^2 + |u(-R)|^2) = 0$, that is, $u(R) = u(-R) = 0$. Note that the variational formulation $\int_{-R}^{R} \left(u'\overline{v'} - k^2 u\overline{v} \right) dx$ directly implies that $u \in H^2(-R, R)$ and that $u'' - k^2 u = 0$. Integrating again by parts in the variational formulation gives

$$\int_{-R}^{R} \left(-u'' - k^2 u \right) \overline{v}\, dx + \left(u'(R) - iku(R) \right)\overline{v(R)}$$
$$- \left(u'(-R) + iku(-R) \right)\overline{v(-R)} = 0 \quad \text{for all } v \in H^1(-R, R).$$

We conclude that $u'(R) - iku(R) = 0$, $u'(-R) + iku(-R) = 0$. Further, $u(R) = u(-R) = 0$ implies that $u'(R) = u'(-R) = 0$, too. The Picard–Lindelöf Theorem yields that u vanishes entirely, and in consequence the variational problem (4.4) is solvable for any wave number $k > 0$.

The last integration by parts also shows that, for general $f \in L^2(-R, R)$, a solution u to the variational problem satisfies

$$\int_{-R}^{R} \left(-u'' - k^2 u + f \right) \overline{v}\, dx + \left(u'(R) - iku(R) \right)\overline{v(R)}$$
$$- \left(u'(-R) + iku(-R) \right)\overline{v(-R)} = 0 \quad \text{for all } v \in H^1(-R, R),$$

especially, the differential equation $u'' + k^2 u = f$ is satisfied in L^2 sense. Finally, we note that the extension u_E of such u to a function on the entire real line,

$$u_E = \begin{cases} u(R)\exp(ik(x - R)), & x > R, \\ u, & -R < x < R, \\ u(-R)\exp(-ik(x + R)), & x < -R, \end{cases}$$

belongs to $H^2_{\text{loc}}(\mathbb{R})$ since (point-)traces and derivatives are continuous at $\pm R$.

4.1.2   Scattering from Bounded Media

In this section, we transfer the concept of the one-dimensional radiation condition (4.2) to higher dimension $m = 2$ or $m = 3$. Our aim is to find a tool that tells whether or not a solution to the Helmholtz equation (4.1) is "outgoing". We need some new notation: For $0 \neq x \in \mathbb{R}^m$ we set $\hat{x} = x/|x|$. The vector $\hat{x} \in \mathbb{S} = \{x \in \mathbb{R}^m : |x| = 1\}$ is a direction in the unit sphere $\mathbb{S}$ of $\mathbb{R}^m$. Assume, for simplicity, that $u \in C^2(\mathbb{R}^m \setminus \overline{B_R})$ is a classical solution to the Helmholtz equation $\Delta u + k^2 u = 0$ in the exterior of the ball $B_R = \{x \in \mathbb{R}^m : |x| < R\}$, $R > 0$. From the 1D case, it seems somehow advisable to consider the restriction of u to the ray $\{r\theta : r \in (R, \infty)\}$ of direction $\theta \in \mathbb{S}$. If we want to check a criterion similar to (4.1), we need to introduce the directional derivative

$$\partial_r u(x) := \hat{x} \cdot \nabla u(x), \quad \text{for } 0 \neq x = r\hat{x}.$$

Then the first condition of (4.1) motivates to consider $\partial_r u(x) - iku(x)$ for $x = r\theta$ and $r > R$. In the 1D case we imposed that the analogue quantity vanishes – this is not meaningful for the multi-dimensional case, since there are more than just two directions. Instead, we prescribe that $\partial_r u(r\theta) - iku(r\theta)$ tends to zero as $r \to \infty$. More precisely, we call u a radiating solution to the Helmholtz equation (or simply *radiating*), if

$$\lim_{r \to \infty} r^{(m-1)/2} \left(\partial_r u - iku\right) = 0, \tag{4.6}$$

uniformly in $\hat{x} \in \mathbb{S}$. This is the so-called Sommerfeld radiation condition. The role of the power $(m-1)/2$ is of course not clear at this point. However, the following consequence shows its meaning: By Cauchy-Schwartz's inequality,

$$\int_{\partial B_R} |\partial_r u - iku|^2 \, ds \leq R^{1-m} \int_{\partial B_R} ds \underbrace{\max_{\partial B_R} R^{m-1} |\partial_r u - iku|^2}_{\to 0 \text{ by } (4.6)} \to 0 \text{ as } R \to \infty$$

because $R^{1-m} \int_{\partial B_R} ds = 2\pi^{m/2}/\Gamma(m/2)$ is uniformly bounded as $R \to \infty$. In analogy to the 1D case we studied in the last section, Sommerfeld's radiation condition guarantees uniqueness of solution to exterior scattering problems on unbounded domains, see for instance the proof Theorem 4.1.4 below.

With Sommerfeld's radiation condition at hand we can now formulate a typical scattering problem: Consider a plane wave $u^i(x) = \exp(ik\,x \cdot \theta)$, $x \in \mathbb{R}^m$, of direction $\theta \in \mathbb{S}$ which is scattered by an inhomogeneous medium described by a refractive index $n^2 : \mathbb{R}^m \to \mathbb{C}$. We assume that $\text{supp}(n^2 - 1) = \overline{D}$ where D is a bounded Lipschitz domain. The presence of the scatterer D creates a scattered field u^s such that the total field $u = u^i + u^s$ satisfies the Helmholtz equation

$$\Delta u + k^2 n^2 u = 0 \quad \text{in } \mathbb{R}^m.$$

Since the scattered field is created locally in D, u^s needs to be an outgoing time-harmonic wave and we impose the Sommerfeld radiation condition (4.6) on u^s. Then, with a little bit of algebra, we find that u^s solves the following problem:

$$\Delta u^s + k^2 n^2 u^s = -k^2 (n^2 - 1) u^i \quad \text{in } \mathbb{R}^m,$$

$$\lim_{r \to \infty} r^{(m-1)/2} \left(\partial_r u - iku \right) = 0 \quad \text{uniformly in } \hat{x} \in \mathbb{S}. \quad (4.7)$$

Later on, we will show solvability of this problem using the radiating fundamental solution of the Helmholtz equation. Here we restrict ourselves to dimension $m = 3$; for $m = 2$, the fundamental solution involves Bessel functions and everything becomes a bit more complicated. The function

$$G_a(x) = \frac{\cos(k|x|) + a \sin(k|x|)}{4\pi|x|} \quad \text{for } x \in \mathbb{R}^3 \setminus \{0\}$$

is a fundamental solution of the Helmholtz equation for all $a \in \mathbb{C}$. This means that (again, in the distributional sense)

$$\Delta G_a + k^2 G_a = -\delta_0, \quad (4.8)$$

where δ_0 is the Dirac distribution at the origin: $\delta_0(\psi) = \psi(0)$ for $\psi \in C_0^\infty(\mathbb{R}^3)$. Recall that (4.8) means that

$$\int_{\mathbb{R}^m} G_a \left(\Delta \psi + k^2 \psi \right) dx = -\psi(0).$$

For a proof of this identity we refer to [39, Theorem 9.4]. At least formally, the last equation means that all functions $u_a(x) = \int_D G_a(x - y) f(y) \, dx$ satisfy $\Delta u + k^2 u = -f$. Again, this statement can be made mathematically rigourous in a distributional sense. Our aim is now to determine $a \in \mathbb{C}$ such that G_a is a radiating solution to the Helmholtz equation, in the sense of (4.6). Since G_a is a radial function, we only consider $x = (r, 0, 0)$, $r > 0$:

$$\partial_r G_a(x) - ik G_a(x)$$
$$= \frac{(a - i)k \cos(kr) - k(1 + ia) \sin(kr)}{4\pi r} - \frac{\cos(kr) + a \sin(kr)}{(4\pi r)^2}.$$

All terms in $O(1/r)$ cancel if and only if $a = i$, thus, we see that precisely for $a = i$

$$\lim_{r \to \infty} r \left(\partial_r G_i(r\theta) - ik G_i(r\theta) \right) = 0$$

uniformly in $\theta \in \mathbb{S}$, that is, G_i satisfies the Sommerfeld radiation condition. In the literature on scattering theory, it is common to define

$$\Phi(x) := G_i(x) = \frac{e^{ik|x|}}{4\pi|x|}.$$

Let now $y \in \mathbb{R}^3$. Then $x \mapsto \Phi(x - y)$ is a translated fundamental solution, which is called a point source at y. Since the limit in (4.6) is requested to be uniform in the angular variable, it is not difficult to see that a translation of a radiating function is still a radiating function. Thus, $x \mapsto \Phi(x - y)$ also satisfies the Sommerfeld radiation condition. Since $(|x - y| - |x|)/(|x - y| + |x|) \le |y|/|x|$, the expansion

$$|x - y| = |x| - \hat{x} \cdot y + \frac{|x - y| - |x|}{|x - y| + |x|} \hat{x} \cdot y + \frac{|y|^2}{|x - y| + |x|}$$

shows that

$$\Phi(x - y) = \Phi(x)\left(e^{-\mathrm{i}k\,\hat{x} \cdot y} + O\left(|x|^{-1}\right) \right), \qquad \text{as } |x| \to \infty.$$

This implies that $\Phi(x - y)$ behaves, for large x, as a radiating spherical wave $\Phi(x)$ with amplitude

$$\Phi_\infty(\hat{x}, y) := \exp(-\mathrm{i}k\,\hat{x} \cdot y), \qquad \hat{x} \in \mathbb{S}, \, y \in \mathbb{R}^m. \tag{4.9}$$

This observation is a special case of the following general theorem, see [19, Theorem 2.5].

Theorem 4.1.1. *Every radiating solution $u \in C^2(\mathbb{R}^3 \backslash \overline{B_R})$ to the Helmholtz equation has the asymptotic behaviour of an outgoing spherical wave,*

$$u(x) = \Phi(x)\left(u_\infty(\hat{x}) + O\left(|x|^{-1}\right) \right) \qquad \text{as } |x| \to \infty.$$

The function $u_\infty \in L^2(\mathbb{S})$ is called the far field pattern of u.

Part (a) of the following lemma is known as Rellich's lemma, and parts (b) and (c) are consequences of that lemma (actually, in dimension $m = 2$ and 3).

Lemma 4.1.2. *Assume that D is a bounded Lipschitz domain with connected complement and let $u \in H^2_{loc}(\mathbb{R}^m \setminus D)$ be a radiating solution of the Helmholtz equation $\Delta u + k^2 u = 0$ in $\mathbb{R}^m \setminus \overline{D}$ with far field u_∞.*

(a) *If $\lim_{r \to \infty} \int_{\partial B_r} |u|^2 \, ds = 0$, then $u = 0$ in the exterior of D.*

(b) *If $\Im\left(\int_{\partial D} u \partial_\nu \bar{u} \, ds \right) \ge 0$, then $u = 0$ in the exterior of D.*

(c) *If $u_\infty = 0$, then $u = 0$ in the exterior of D.*

A proof, in the context of spaces of continuously differentiable functions, can be found in [19, Theorems 2.12 & 2.13]. In a more general context, the result is shown in [39, Theorem 9.6]. Note that the Sommerfeld radiation condition is well defined for a solution of the Helmholtz equation in $H^2_{loc}(\mathbb{R}^m \setminus D)$ since such a solution is a smooth function by interior elliptic regularity results, see [39].

The Lippmann–Schwinger Integral Equation

The scattering problem (4.7) can be reformulated as a volumetric integral equation posed on the scattering object D. This so-called Lippmann–Schwinger integral equation admits a comparatively simple solution theory. The integral equation approach for the medium scattering problem is based on the fundamental solution for the Helmholtz equation. In the last section, we introduced this function for dimension $m = 3$ and we restrict ourselves to three dimensions here, too. In two dimensions, the theory works very much in the same way but the fundamental solution involves Hankel functions, which we want to avoid for simplicity. Thus, we recall that $\Phi(x, y) = e^{ik|x-y|}/(4\pi|x - y|)$ for $x \neq y \in \mathbb{R}^3$, and formally define the volume potential of a function $f : D \mapsto \mathbb{C}$ by

$$V(f) = \int_D \Phi(\cdot, y) f(y) \, dy.$$

In the following, it is convenient to extend functions in $L^2(D)$ implicitly by zero to all of $\mathbb{R}^3$. The analysis of medium scattering is greatly simplified by means of the following proposition.

Proposition 4.1.3. *The volume potential V is a bounded operator from $L^2(D)$ into $H^2(B)$ for all balls $B \subset \mathbb{R}^3$. For $f \in L^2(D)$, the function $v = V(f) \in H^2_{\mathrm{loc}}(\mathbb{R}^3)$ solves $\Delta v + k^2 v = -f$ in $L^2(\mathbb{R}^3)$.*

For a proof of this proposition we refer to [19, Theorem 8.3], and, again in a more general context, to [39, Chapter 6]. The solution u^s to the scattering problem (4.7) solves $\Delta u^s + k^2(1 + q)u^s = -k^2 q u^i$ subject to the Sommerfeld radiation condition. We are going to tackle this problem in greater generality and find a solution $v \in H^2_{\mathrm{loc}}(\mathbb{R}^3)$ to

$$\Delta v + k^2(1 + q)v = -k^2 q f \quad \text{in } \mathbb{R}^3 \quad \text{for } f \in L^2(D), \tag{4.10}$$

where v also satisfies the Sommerfeld radiation condition (4.6). The trick is to seek a solution in form of a volume potential $V(g)$ with unknown density $g \in L^2(D)$. Note that the striking advantage of this approach is that the radiation condition is automatically built in: Since $\Phi(\cdot, y)$ satisfies Sommerfeld's radiation condition, $v := V(g)$ is by linearity a radiating function for any $g \in L^2(D)$. Since $\Delta v + k^2 v = -g$ and since we want to solve $\Delta v + k^2 v = -k^2 q(f + v)$, the density g needs to be equal to $k^2 q(f + v)$. Plugging in this equation in the definition of v yields $v - k^2 V(qv) = k^2 V(qf)$ in $L^2(\mathbb{R}^3)$. We restrict this equation to D since $f \in L^2(D)$ and knowledge of $v|_D$ is sufficient to reconstruct V everywhere in $\mathbb{R}^3$ due to $v = k^2 V(q(f + v))$. Thereby we obtain the Lippmann–Schwinger integral equation

$$v - k^2 V(qv) = k^2 V(qf) \tag{4.11}$$

which is an operator equation in $L^2(D)$.

Theorem 4.1.4. *Let $q \in L^\infty(D)$ such that $\Im(q) \geq 0$ and $f \in L^2(D)$.*

(a) *If a function $v \in H^2_{\mathrm{loc}}(\mathbb{R}^3)$ solves the Helmholtz equation (4.10) in the weak sense and the radiation condition (4.6), then the restriction $v|_D$ solves the Lippmann–Schwinger integral equation (4.11). If $v \in L^2(D)$ solves the Lippmann–Schwinger integral equation, then $k^2 V(q(v+f))$ provides an extension of v to a function in $H^2_{\mathrm{loc}}(\mathbb{R}^3)$ solving (4.10) together with (4.6).*

(b) *The Lippmann–Schwinger equation (4.11) has a unique solution for all $k > 0$ and $f \in L^2(D)$, and this solution depends continuously on f.*

Proof. (a) Assume first that $v \in H^2_{\mathrm{loc}}(\mathbb{R}^3)$ is a radiating solution to (4.10). The volume potential $V(k^2 q(f + v))$ satisfies

$$\Delta V(k^2 q(v + f)) + k^2 V(k^2 q(v + f)) = -k^2 q(v + f)$$

and hence $w = v - V(k^2 q(v + f))$ is a solution of the Helmholtz equation $\Delta w + k^2 w = 0$ with zero right hand side. Moreover, w satisfies the Sommerfeld radiation condition and is hence an entire radiating solution of the Helmholtz equation. It is well known that such a function vanishes identically in all of $\mathbb{R}^3$, see [19].

Assume now that $v \in L^2(D)$ solves the Lippmann–Schwinger equation (4.11). We extend v by the volume potential $k^2 V(q(f + v))$ to a radiating solution of the Helmholtz equation in $\mathbb{R}^3$. As mentioned above, V is a bounded operator from $L^2(D)$ into $H^2_{\mathrm{loc}}(\mathbb{R}^3)$, thus, $v \in H^2_{\mathrm{loc}}(\mathbb{R}^3)$. We computed in the first part of the proof that $\Delta V(k^2 q(v + f)) + k^2 V(k^2 q(v + f)) = -k^2 q(v + f)$ in $\mathbb{R}^3$. From the latter equation combined with the Lippmann–Schwinger equation it follows that $\Delta v + k^2 v = -k^2 q(v + f)$, which means that $\Delta v + k^2(1 + q)v = -k^2 q f$. The proof of this part is complete.

(b) The boundedness of V from $L^2(D)$ into $H^2_{\mathrm{loc}}(\mathbb{R}^3)$ implies that $f \mapsto k^2 V(qf)|_D$ is a compact operation on $L^2(D)$. From the Lippmann–Schwinger equation (4.11) one observes that Riesz theory [34] (one could also use Fredholm theory here) implies that the Lippmann–Schwinger equation is solvable for any right-hand side if its only solution for $f = 0$ is the trivial solution. Hence, to finish the proof we show injectivity of $I - k^2 V(q \cdot)|_D$ on $L^2(D)$.

Assume that $v - k^2 V(qv)|_D = 0$ for $v \in L^2(D)$. According to part (a) of the proof, the extension of v by $k^2 V(qv)$ to all of $\mathbb{R}^3$ is a radiating solution of (4.11) for zero right hand side. The mapping properties of V imply that $v \in H^2_{\mathrm{loc}}(\mathbb{R}^3)$. Using this regularity result, we conclude by Green's first identity in $B_r := \{x \in \mathbb{R}^3 : |x| < r\}$, $r > 0$ large enough, that

$$0 = -\int_{B_r} (\Delta v + k^2 n^2 v)\overline{v}\, dx = \int_{B_r} (|\nabla v|^2 - k^2 n^2 |v|^2)\, dx - \int_{\partial B_r} \overline{v}\partial_r v\, ds$$

which implies

$$\Im\left(\int_{\partial B_r} v\partial_r \overline{v}\, ds\right) = k^2 \int_{B_r} \Im(n^2)|v|^2\, dx \geq 0$$

and Theorem 4.1.2 (c) implies that v vanishes in the exterior of B_r for r large enough. However, in this situation the unique continuation property [31, Lemma 5.3 & Theorem 5.4] implies that v vanishes in all of $\mathbb{R}^3$.   □

In two dimensions, the kernel of the Lippmann–Schwinger equation involves Bessel functions – apart from this technical complication the last result transfers one-to-one to two dimensions.

A particularly nice feature of the integral equation approach to solve scattering problems is that the far field of the solution can be easily computed. Since $v = k^2 V(q(f + v))$, the far field of v can be represented as the far field of a volume potential with density $k^2 q(f + v)$. By linearity, the far field pattern of the volume potential again takes the form of an integral operator where the kernel Φ is replaced by the far field Φ_∞ of the kernel: For $g \in L^2(D)$,

$$(V(g))_\infty = \left(\int_D \Phi(\cdot, y) g(y) \, dy \right)_\infty = \int_D \Phi_\infty(\cdot, y) g(y) \, dy.$$

We computed above that $\Phi_\infty(\hat{x}, y) = \exp(-ik\,\hat{x}\cdot y)$, thus, the far field of a solution to (4.11) is given by

$$v_\infty(\hat{x}) = k^2 \left(V(q(v + f)) \right)_\infty (\hat{x}) = k^2 \int_D e^{-ik\,\hat{x}\cdot y} q(y)(v(y) + f(y)) \, dy, \qquad \hat{x} \in \mathbb{S}.$$

Apart from existence theory, the Lippmann–Schwinger equation can also be used for numerical purposes [49]. One exploits that the volume potential is a convolution operator and hence a multiplication operator in the Fourier domain. Using fast Fourier transform, this allows to rapidly evaluate the operator arising in the Lippmann–Schwinger integral equation. The discrete system is then solved using an iterative method. Details on this spectral method can be found in [49, 46]. For a two dimensional computational example, we choose a kite-shaped obstacle $D = (-1, 1)^2 \setminus [0, 1) \times (-1, 0]$ in which the contrast q equals two. The wave number is $k = 3\pi$, the wave length $\lambda = 2\pi/k$ is hence $2/3$, and the direction of the incident plane wave is $d = (-1/\sqrt{2}, 1/\sqrt{2})^\top$. Figure 4.2 shows the obstacle together with the real parts of total, incident, and scattered field. The plot of the scattered field illustrates the Sommerfeld radiation condition: Away from the scatterer the field behaves like a spherical wave with a certain amplitude.

4.1.3 Scattering from Periodic Media

In the last section we studied scattering of an incident time-harmonic wave at a bounded inhomogeneity. Now, we will investigate scattering of waves at periodic structures. For simplicity, we start in two dimensions and assume that the structure's period in the lateral variable x_1 is 2π. Again, our model for time-harmonic wave propagation is the Helmholtz equation $\Delta u + k^2 n^2 u = 0$, where we assume that the periodic index of refraction $n^2 \in L^\infty(\mathbb{R}^2)$ with $\Re(n^2) \geq c > 0$, $\Im(n^2) \geq 0$,

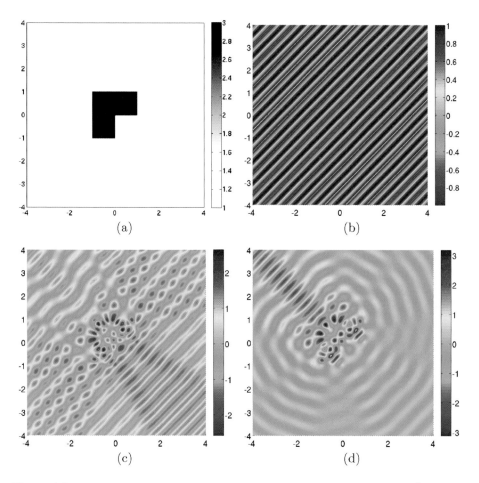

Figure 4.2: Scattering from an L-shaped inhomogeneous medium $D = (-1,1)^2 \setminus [0,1) \times (-1,0]$ where the contrast is equal to two. The wave number k equals 3π, the wave length λ is $2/3$. (a) Refractive index n^2 (b) Real part of incident field $d = (-1/\sqrt{2}, 1/\sqrt{2})^\top$ (c) Real part of total field (d) Real part of scattered field.

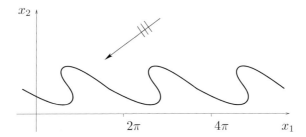

Figure 4.3: A periodic structure (e. g., a surface or an interface) is hit by a plane wave under a certain angle of incidence. In this section, we consider the problem of finding the associated scattered field.

equals one outside the periodic penetrable object. Periodicity of the scattering structure thus means that $n^2(x_1 + 2\pi, x_2) = n^2(x_1, x_2)$, $x = (x_1, x_2)^\top$ in $\mathbb{R}^2$. We will further assume that $n^2(x_1, x_2)$ equals one for $|x_2|$ large enough. Our main interest lies in solving the problem of scattering of a downwards traveling plane wave

$$u^i(x) = e^{ik\,x\cdot d} = e^{ik(x_1 d_1 + x_2 d_2)}, \qquad d \in \mathbb{S}, \, d_2 < 0$$

at the periodic structure, compare Figure 4.3. This problem arises for instance when one considers scattering of electromagnetic waves from a dielectric three-dimensional structure which is independent of one coordinate, when the wave vector indicating the direction of the wave is orthogonal to the invariance axis of the structures, see, e. g., [23, 12]. These special structures are called *diffraction gratings*, and that is why the problem we consider in this section is sometimes called the diffraction grating problem.

   A central difference to the problem from the last section is that the Sommerfeld radiation condition is not the physically adequate radiation condition for the periodic scattering problem. Indeed, Sommerfeld's radiation condition treats all directions in the same way, but solutions to the periodic problem should behave differently in the vertical periodic layer compared to the homogeneous half-spaces above and below. As in the last section, we would like to impose that the scattered field propagates away from the structure, which means that the scattered field above and below the structure should radiate upwards and downwards, respectively. We formulate this condition with the help of Fourier analysis.

   First, we note that the incident field u^i does not share 2π–periodicity of the structure,

$$u^i(x_1 + 2\pi, x_2) = e^{2\pi i k d_1} u^i(x_1, x_2) = e^{2\pi i \alpha} u^i(x_1, x_2) \tag{4.12}$$

where we introduced the *quasiperiodicity* $\alpha := k d_1$. We call functions that satisfy (4.12) α-*quasiperiodic* with respect to x_1. For an α-quasiperiodic function, a translation by 2π in x_1 causes a phase shift by $\exp(2\pi i \alpha)$. Since the refractive index n^2 is assumed to be 2π periodic, it seems from a physical point of view meaningful to assume that u^s is α-quasiperiodic, too. This assumption is intimately connected with uniqueness of solution for this scattering problem: From the equation $\Delta u + k^2 n^2 u = 0$ for the total field, we note that $\Delta u^s + k^2 n^2 u^s = -k^2 q u^i$ with $q = n^2 - 1$. A translation by 2π in x_1 direction of the right hand side hence creates a factor $\exp(2\pi i\alpha)$, whereas the differential operator on the right in invariant under such a shift (because $n^2(x_1 + 2\pi, x_2) = n^2(x_1, x_2)$). Hence, we see that for a given solution u^s also $\exp(2\pi i\alpha)u^s(x_1, x_2)$ is a solution, and if u^s was *not* α-quasiperiodic, this procedure would generate a new solution.

The 2π periodic function $x_1 \mapsto e^{-i\alpha x_1}u^s(x_1, x_2)$ can be written as a Fourier series,

$$e^{-i\alpha x_1}u^s(x_1, x_2) = \sum_{j\in\mathbb{Z}} a_j(x_2)e^{ijx_1}.$$

Assume that h is so large that $h > \sup\{|x_2| : (x_1, x_2) \in \operatorname{supp}(q)\}$. Then $\Delta u^s + k^2 u^s = 0$ in the half space $\{(x_1, x_2) : x_2 > h\}$ and plugging our ansatz (formally) into this equation we get

$$\sum_{j\in\mathbb{Z}}(a_j'' + a_j(x_2)(k^2 - (j+\alpha)^2))e^{i(j+\alpha)x_1} = 0.$$

Thus, $a_j^\pm(x_2) = \hat{u}_j \exp(\pm i\sqrt{k^2 - (j+\alpha)^2}\,(x_2 \mp h))$, where we define the square root via holomorphic extension from the positive real numbers to the complex plane slit at the negative imaginary axis. The numbers $\hat{u}_j$ will play the role of constants that we will later on identify as Fourier coefficients of u^s. We have to choose the sign in front of the root appropriately to obtain an upwards or a downwards propagating solution. At this point we recall what we found out in the analysis of the one-dimensional situation in Section 4.1.1: For $k^2 > (j+\alpha)^2$, a_j^+ propagates upwards, whereas a_j^- propagates downwards. Further, for $k^2 < (j+\alpha)^2$, a_j^+ decays exponentially as $x_2 \to \infty$ – we call such modes evanescent – and a_j^- decays exponentially as $x_2 \to -\infty$. Since we seek for bounded solutions (physically corresponding to finite-energy solutions), we hence prescribe that

$$u^s(x) = \sum_{j\in\mathbb{Z}} \hat{u}_j^+ e^{i(\alpha_j x_1 + \beta_j(x_2-h))} \quad \text{for } x_2 \geq h, \quad \alpha_j := j + \alpha,\ \beta_j = \sqrt{k^2 - \alpha_j^2},$$

$$u^s(x) = \sum_{j\in\mathbb{Z}} \hat{u}_j^- e^{i(\alpha_j x_1 - \beta_j(x_2+h))} \quad \text{for } x_2 \leq -h, \quad (4.13)$$

and we require uniform convergence of the two series in the two half-spaces. This is the famous Rayleigh expansion condition [44]. Note that $k^2 > (j+\alpha)^2$ implies that the j-th mode $\exp(i(\alpha_j x_1 - \beta_j(x_2-h)))$ is a propagating mode, whereas $k^2 < (j+\alpha)^2$ implies that $\exp(i(\alpha_j x_1 - \beta_j(x_2 - h)))$ is an evanescent mode. Finally, we note that the numbers $\hat{u}_j^\pm$ are the Fourier(–Rayleigh) coefficients of the restriction of u^s to the lines

$$\Gamma_{\pm h} = \{(x_1, \pm h) : -\pi < x_1 < \pi\},$$

that is,

$$\hat{u}_j^\pm = \frac{1}{2\pi} \int_{\Gamma_{\pm h}} u^s(x) e^{-i(j+\alpha)x_1} \, ds = \frac{1}{2\pi} \int_{-\pi}^{\pi} u^s(x_1, \pm h) e^{-i(j+\alpha)x_1} \, d(x_1). \quad (4.14)$$

Variational Formulation and Existence of Solution

We have now prepared all tools to precisely formulate the periodic medium scattering problem, to derive a variational formulation, and to determine conditions such that this formulation is uniquely solvable. We do all this under the general (technical) assumption that

$$k^2 \neq \alpha_j^2 \qquad \text{for all } j \in \mathbb{Z}.$$

This assumption appears a bit from nowhere and a few comments are in order. Later on, we shall see that this condition means that k^2 does not correspond to a so-called Rayleigh(–Wood) frequency at which the number of propagating modes changes. Also, Rayleigh frequencies might involve nonuniqueness phenomena, thus, roughly speaking, the latter condition is a nonresonance condition. Taking a purely technical point of view, if $k^2 = \alpha_j^2$ our later (simple) analysis breaks down, since several times we are forced to divide by this term.

For an α-quasiperiodic incident plane wave u^i such that $\Delta u^i + k^2 u^i = 0$ in $\mathbb{R}^2$ we seek an α-quasiperiodic function $u^s \in H^2_{\text{loc}}(\mathbb{R}^2) = \{u \in \mathcal{D}'(\mathbb{R}^2) : u|_B \in H^2(B)$ for all open balls $B \subset \mathbb{R}^2\}$ that satisfies $\Delta u^s + k^2 n^2 u^s = -k^2 q u^i$ in the weak sense and further the Rayleigh expansion conditions (4.13).

Due to periodicity, we can restrict ourselves to the strip $\Omega := \{(x_1, x_2) : -\pi < x_1 < \pi\}$ and further will start to seek a solution in

$$H^1_{\alpha,\text{loc}}(\Omega) := \{u \in H^1_{\text{loc}}(\Omega) : u = U|_{\Omega_h} \text{ for some } \alpha\text{-quasiperiodic } U \in H^1_{\text{loc}}(\mathbb{R}^2)\}.$$

Further, a variational formulation of the scattering problem should be defined in a bounded domain, say, in $\Omega_h := (-\pi, \pi) \times (-h, h)$, and thus we seek a solution in

$$H^1_\alpha(\Omega_h) := \{u \in H^1(\Omega_h) : u = U|_\Omega \text{ for some } U \in H^1_{\alpha,\text{loc}}(\Omega)\}. \quad (4.15)$$

The Rayleigh expansion condition has now to be enforced by the variational formulation, since it is not included in the space $H^1_\alpha(\Omega_h)$. Assume that $u^s \in H^2_{\text{loc}}(\mathbb{R}^2)$ is a solution to our quasiperiodic scattering problem. By multiplication of $\Delta u^s + k^2 n^2 u = -k^2 q u^i$ by a test function $\overline{v}$, $v \in H^1_\alpha(\Omega_h)$, and by using Green's identity in Ω_h, we find that

$$\int_{\Omega_h} (\nabla u^s \cdot \nabla \overline{v} - k^2 n^2 u^s \overline{v}) \, ds - \int_{\partial \Omega_h} \overline{v} \, \partial_\nu u^s \, ds = k^2 \int_{\Omega_h} q u^i \overline{v} \, dx.$$

Here, ν is the exterior normal field (defined almost everywhere) to Ω_h and $\partial_\nu := \nu \cdot \nabla$ the the normal derivative. Due to $\overline{v}(-\pi, \cdot) = \exp(-2\pi i \alpha)\overline{v}(\pi, \cdot)$ and because the direction of the exterior normal field ν changes, we have $\partial_\nu u^s(-\pi, \cdot) =$

$-\exp(2\pi i\alpha)\partial_\nu u^s(\pi,\cdot)$ (here, we exploited that $u^s \in H^2_{\text{loc}}(\mathbb{R}^2)$). Thus, we compute that

$$\int_{\partial\Omega_h} \bar{v}\,\partial_\nu u^s\,ds = \int_{\Gamma_h} \bar{v}\,\partial_\nu u^s\,ds + \int_{\Gamma_{-h}} \bar{v}\,\partial_\nu u^s\,ds$$

$$= \int_{\Gamma_h} \bar{v}\,\partial_2 u^s\,ds - \int_{\Gamma_{-h}} \bar{v}\,\partial_2 u^s\,ds.$$

Further, from the Rayleigh expansion condition (4.13) we formally see that

$$\partial_2 u^s|_{\Gamma_h} = i\sum_{j\in\mathbb{Z}} \beta_j \hat{u}_j^+ e^{i\alpha_j x_1} =: T^+(u^s|_{\Gamma_h}), \qquad \text{and}$$

$$-\partial_2 u^s|_{\Gamma_{-h}} = i\sum_{j\in\mathbb{Z}} \beta_j \hat{u}_j^- e^{i\alpha_j x_1} =: T^-(u^s|_{\Gamma_{-h}}). \qquad (4.16)$$

The two *Dirichlet-to-Neumann* operators

$$T^\pm(\phi^\pm) = i\sum_{j\in\mathbb{Z}} \beta_j \hat{\phi}_j^\pm e^{\alpha_j x_1}, \qquad \hat{\phi}_j^\pm = \frac{1}{2\pi}\int_{-\pi}^{\pi} \phi^\pm(x_1) e^{-i(j+\alpha)x_1}\,d(x_1),$$

are bounded from $H_\alpha^{1/2}(\Gamma_{\pm h})$ into $H_\alpha^{-1/2}(\Gamma_{\pm h})$, where

$$\|\phi^\pm\|_{H_\alpha^{-1/2}(\Gamma_{\pm h})}^2 = \sum_{j\in\mathbb{Z}} |k^2 - \alpha_j^2|^{-1/2} |\hat{\phi}_j^\pm|^2,$$

and $\hat{\phi}_j^\pm$ are the Fourier coefficients of $\phi^\pm \in H_\alpha^{-1/2}(\Gamma_{\pm h})$. Indeed,

$$\|T^\pm(\phi^\pm)\|_{H_\alpha^{-1/2}(\Gamma_{\pm h})}^2 = \sum_{j\in\mathbb{Z}} |k^2 - \alpha_j^2|^{-1/2} |\beta_j|^2 |\hat{\phi}_j^\pm|^2$$

$$= \sum_{j\in\mathbb{Z}} |k^2 - \alpha_j^2|^{1/2} |\hat{\phi}_j^\pm|^2 = \|\phi^\pm\|_{H_\alpha^{1/2}(\Gamma_{\pm h})}.$$

Note that it is our general assumption that $k^2 \neq \alpha_j^2$ for all $j \in \mathbb{Z}$ which yields that the expression defining $\|\cdot\|_{H_\alpha^{-1/2}(\Gamma_{\pm h})}$ is positive definite and hence defines a norm. We replace the term $\int_{\Gamma_h} \bar{v}\,\partial_\nu u^s\,ds$ in (4.16) by $\int_{\Gamma_h} \bar{v}\,T^+(u^s)\,ds$, proceed analogously with the term on Γ_{-h}, and note that the last integral is well defined for $u^s, v \in H_\alpha^1(\Omega_h)$. Indeed, it can be shown that $H_\alpha^{1/2}(\Gamma_{\pm h})$ is the trace space of $H_\alpha^1(\Omega_h)$ on $\Gamma_{\pm h}$ (see [39, 34] for more on Sobolev spaces), and that there holds

$$\left| \int_{\Gamma_h} \bar{v}\,T^+(u^s|_{\Gamma_h})\,ds \right| \leq \|v\|_{H_\alpha^{1/2}(\Gamma_h)} \|T^+(u^s)\|_{H_\alpha^{-1/2}(\Gamma_h)}$$

$$\leq C\|v\|_{H_\alpha^1(\Omega_h)} \|u^s\|_{H_\alpha^{1/2}(\Gamma_h)} \leq C\|v\|_{H_\alpha^1(\Omega_h)} \|u^s\|_{H_\alpha^1(\Omega_h)}. \qquad (4.17)$$

We hence obtain the following variational formulation, where we replace the incident field u^i for convenience by an arbitrary function $f \in L^2(\Omega_h)$: Find $u^s \in H^1_\alpha(\Omega_h)$ such that for all $v \in H^1_\alpha(\Omega_h)$ it holds

$$\mathcal{B}_{\mathrm{p}}(u^s, v) := \int_{\Omega_h} (\nabla u^s \cdot \nabla \overline{v} - k^2 n^2 u^s \overline{v})\, ds$$
$$- \int_{\Gamma_h} \overline{v}\, T^+(u^s)\, ds - \int_{\Gamma_h} \overline{v}\, T^-(u^s)\, ds = k^2 \int_{\Omega_h} q f \overline{v}\, dx. \quad (4.18)$$

With the help of (4.17) it is easy to see that $\mathcal{B}_{\mathrm{p}}$ is a bounded sesquilinear form on $H^1_\alpha(\Omega_h)$. We note a couple of further estimates for this form: First, using the Plancherel identity,

$$-\Re\left(\int_{\Gamma_{\pm h}} \overline{u^s} T^\pm(u^s)\, ds \right) = -\Re(\mathrm{i} \sum_{j \in \mathbb{Z}} \beta_j |\hat{u}_j^\pm|^2)$$
$$= \sum_{j\,:\,\alpha_j^2 > k^2} \sqrt{\alpha_j^2 - k^2} |\hat{u}_j^\pm|^2 \geq 0,$$

and for the imaginary part we find

$$-\Im\left(\int_{\Gamma_{\pm h}} \overline{u^s}\, T^\pm(u^s|_{\Gamma_h})\, ds \right) = - \sum_{j\,:\,\alpha_j^2 < k^2} \sqrt{k^2 - \alpha_j^2} |\hat{u}_j^\pm|^2. \quad (4.19)$$

For the sesquilinear form this implies that

$$\Re\left(\mathcal{B}_{\mathrm{p}}(u^s, u^s) \right) \geq \|u\|_{H^1_\alpha(\Omega_h)}^2 + \int_{\Omega_h} (1 - k^2 n^2) |u^s|^2\, dx,$$

and since $L^2(\Omega_h)$ is compactly embedded in $H^1_\alpha(\Omega_h)$ (this is Rellich's compact embedding lemma in the periodic setting), $\mathcal{B}_{\mathrm{p}}$ satisfies a Gårding inequality. Fredholm theory [39] implies that existence of solution for problem (4.18) follows from uniqueness of solution. If we can show that for $u^i = 0$ the only solution to (4.18) is the trivial solution, then there exists a unique solution for any incident field u^i – actually, this statement then holds for any right hand side $f \in L^2(\Omega_h)$.

So, let us investigate under which conditions uniqueness of solution holds, and assume that $u^s \in H^1_\alpha(\Omega_h)$ satisfies $\mathcal{B}_{\mathrm{p}}(u^s, v) = 0$ for all $v \in H^1_\alpha(\Omega_h)$. Then

$$\Im\left(\int_{\Omega_h} k^2 n^2 |u^s|^2\, ds \right) + \Im\left(\int_{\Gamma_h} \overline{u}\, T^+(u^s)\, ds \right) + \Im\left(\int_{\Gamma_{-h}} \overline{u}\, T^-(u^s)\, ds \right) = 0.$$

Since all three terms are nonnegative (the first one by our assumption $\Im(n^2) \geq 0$, the other two by (4.19)), they all have to vanish. We distinguish now two cases where we can prove uniqueness of solution:

(1) If $\Im(k^2n^2) > 0$ on some open set of Ω_h, then u^s vanishes on this set. From the unique continuation property for elliptic partial differential equations, see, e. g., [48, 31], we obtain that u^s vanishes in Ω_h. Hence, we have uniqueness of solution and thereby existence holds, too. This case corresponds to a medium that is (partially) absorbing. Roughly speaking, the absorbing medium is usually easier to treat than the nonabsorbing medium we consider next.

(2) Assume that $\Im(k^2n^2) = 0$. Due to

$$\Im\left(\int_{\Gamma_{\pm h}} \overline{u^s}T^\pm(u^s)\,ds\right) = \sum_{j:\alpha_j^2<k^2} \sqrt{k^2 - \alpha_j^2}\,|\hat{u}_j^\pm|^2 = 0,$$

u^s consists only of evanescent modes. If u^s is not the trivial solution, we call it a *surface wave* since it is exponentially localised near the inhomogeneous medium. Existence of surface waves is indeed possible (see the next Section 4.1.4), even if this situation is "rare": The form $\mathcal{B}_\mathrm{p} = \mathcal{B}_\mathrm{p}(k)$ depends holomorphically on $k \in \{c \in \mathbb{C} : \Re(c) > 0, c \neq \alpha_j - i\eta, \eta > 0\}$ since the volumetric term is a polynomial in k and the boundary terms depend holomorphically on k. For a precise definition of holomorphic operator-valued functions we refer to [25, 30, 19]. The fact that $\mathcal{B}_\mathrm{p}$ depends holomorphically on k and also satisfies a Gårding inequality remarkably implies the following property [25]: The set of *singular frequencies* $\{k_j\}$ in $\mathbb{R}_+$ where nonuniqueness occurs is at most countable and has no accumulation point other than ∞: For all $0 < k \neq \{k_j\}_{j=1}^N$, $N \in \mathbb{N}\cup\infty$, the variational problem (4.18) is uniquely solvable for any right-hand side.

In the beginning of this section we started to seek for a *strong solution u^s* of the scattering problem in H^2, such that the equation $\Delta u + k^2n^2u = -k^2qu^i$ is satisfied in the L^2 sense. By elliptic regularity results [39] it can be shown that a solution $u^s \in H_\alpha^1(\Omega_h)$ to (4.18) does indeed belong to $H_\alpha^2(\Omega_h)$. Further, it is important to note that (as in the one-dimensional case) the solution u can be extended by the Rayleigh expansion condition (4.13) to a function in $H_{\alpha,\mathrm{loc}}^2(\Omega)$ which solves the Helmholtz equation $\Delta u + k^2(1+q)u = -k^2qf$. This is due to the fact that trace and normal derivative of u and its extension agree on $\Gamma_{\pm h}$. We do not prove the last two results here, but recapitulate the main result of this section.

Theorem 4.1.5.

(1) *If $\Im(k^2n^2) > 0$ on some open subset of Ω_h, then (4.18) is uniquely solvable for any right-hand side $f \in L^2(\Omega_h)$.*

(2) *If $\Im(k^2n^2) = 0$ in Ω_h, then there exists a (possibly finite) sequence $\{k_j\}_{j=1}^N$, $N \in \mathbb{N}\cup\infty$, such that for all $0 < k \neq \{k_j\}_{j=1}^N$, the variational problem (4.18) is uniquely solvable for any right-hand side $f \in L^2(\Omega_h)$. If $N = \infty$, then $k_j \to \infty$ as $j \to \infty$.*

There is of course much more theory on periodic scattering problems than we presented in this section. Interesting further topics include for instance other boundary conditions [43, 30, 12, 23], a-priori estimates and existence theory via Rellich identities [12] improving the above existence theory, analysis of nonunique-ness phenomena [23, 12, 40], electromagnetic waves [1, 47], and numerical analysis [10]. These references are by no means complete but rather might serve as a starting point for readers interested in these topics.

Again, we want to illustrate the last result by a numerical example, which is computed by a finite element method using the software FreeFem++ (see `http://www.freefem.org`). The radiation condition is implemented using a perfectly matched layer, see, e. g., [10]. Figure 4.4 shows two periods of the periodic medium consisting of a layer $\{(x_1, x_2)^\top : f(x_1) - 1 < x_2 < f(x_1) + 1\}$ with a 2π period function f such that $f(x_1) = 3$ for $x_1 \in [-\pi/2, \pi/2)$ and $f(x_1) = 4$ for $x_1 \in [\pi/2, 3\pi/2)$. Further, real parts of incident, total, and scattered field are shown. The refractive index equals 1 in the half spaces above and below the inhomogeneous medium, and 1.5 in the layer itself. The wave number equals 2π, that is, the wave length is 1 in the domain exterior to the layer.

4.1.4   Nonuniqueness for the Periodic Scattering Problem

To complete the study of the periodic scattering problem from the last section, we have a brief look at nonuniqueness phenomena and show the existence of surface waves under certain assumptions on the periodic refractive index. We follow the analysis in [12]; our notation and assumptions are as in the last section, especially, we assume that $k^2 \neq \alpha_j^2$ for all $j \in \mathbb{Z}$. Due to Theorem 4.1.5 we can further assume that the periodic refractive index n^2 is real valued. The main result in this section is that

$$\int_{\Omega_h} (n^2 - 1) \, dx > 0,$$

which implies that there is at least one $k > 0$ such that the homogeneous problem corresponding to (4.18) has a nontrivial solution. Such a nontrivial solution decays exponentially as $x_2 \to \pm\infty$ and is in this sense concentrated in a neighbourhood of the diffraction grating. Therefore, these solutions are often called surface waves.

We first reformulate the nonuniqueness problem as a spectral problem. If $u \in H_\alpha^1(\Omega_h)$ solves $\mathcal{B}_p(u, v) = 0$ for all $v \in H_\alpha^1(\Omega_h)$, then we take the imaginary part of $\mathcal{B}_p(u, u) = 0$ to see from (4.19) that all radiating modes for which $k^2 > \alpha_j^2$ vanish. Thus, the unique extension of u via the Rayleigh expansion condition (4.13) possesses only evanescent modes and therefore this extension decays exponentially as $x_2 \to \pm\infty$. This implies that $u \in H_\alpha^1(\Omega)$ (recall that $\Omega := \{(x_1, x_2) : 0 \leq x_1 \leq 2\pi\}$). Since u is a weak solution to the Helmholtz equation,

$$a(u, v) = \int_\Omega \nabla u \cdot \nabla v \, dx = k^2 \int_\Omega n^2 u \bar{v} \, dx =: k^2 \langle u, v \rangle. \tag{4.20}$$

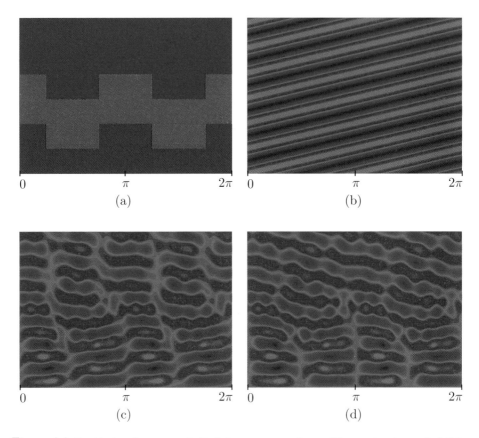

0   π   2π   0   π   2π
                (a)   (b)

0   π   2π   0   π   2π
                (c)   (d)

Figure 4.4: Scattering from a periodic inhomogeneous layer. The contrast equals 0.5 in the layer, the wave number is 2π. (a) Refractive index n^2 (b) Real part of incident field with direction (c) Real part of total field (d) Real part of scattered field.

The inner product $\langle u, v \rangle = \int_\Omega n^2 u\bar{v}\, dx$ is equivalent to the standard inner product on $L^2(\Omega)$ since, by assumption, $0 < c \leq n^2 \leq \|n^2\|_{L^\infty(\Omega)} < \infty$. A norm is then defined by $\|u\|_2 = \langle u, u \rangle^{1/2}$. Now we can interpret (4.20) as eigenvalue problem for A_α defined by $\langle A_\alpha u, v \rangle = a(u, v)$ for all $v \in H^1_\alpha(\Omega)$, with domain

$$D(A_\alpha) = \{u \in H^1_\alpha(\Omega) : \exists C : a(u, v) \leq C\|v\| \text{ for all } v \in H^1_\alpha(\Omega)\}.$$

The eigenvalue problem is to find $(\lambda, u) \in \mathbb{R}_{>0} \times D(A_\alpha)$ such that $A_\alpha u = \lambda u$.

We state the following lemma from [12] without proof.

Lemma 4.1.6.

1. *A_α is a positive selfadjoint operator on $(L^2(\Omega), \langle \cdot, \cdot \rangle)$.*

2. *If k is an eigenvalue of A_α, then (4.18) is not uniquely solvable.*

3. *The continuous spectrum $\sigma_c(A_\alpha)$ of A_α is included in $[\alpha^2, +\infty[$ and the spectrum $\sigma(A_\alpha)$ of A_α is included in $[\alpha^2/\|n^2\|_\infty, \infty[$.*

4. *The following min-max principle for eigenvalues below the continuous spectrum holds: If*

$$\lambda_m(\alpha) = \inf_{V_m \subset H^1_\alpha(\Omega),\, \dim(V_m) = m} \sup_{u \in V_m} \frac{a(u, u)}{\|u\|_2} < \alpha^2, \quad m \in \mathbb{N},$$

then $\lambda_1(\alpha), \ldots, \lambda_m(\alpha)$ are the first m eigenvalues of A_α.

Using the last lemma, it is possible to show that the condition $\int_\Omega(n^2 - 1)\, dx > 0$ implies that A_α possesses an eigenvalue $\lambda > 0$ below the continuous spectrum. Thus, the periodic scattering problem (4.18) is not uniquely solvable for wave number $k = \lambda^{1/2}$.

Theorem 4.1.7. *If $\int_\Omega(n^2 - 1)\, dx > 0$, then there is at least one eigenvalue $\lambda_1(\alpha)$ of A_α for any $\alpha \in (0, 1]$.*

Proof. Set, for $j \in \mathbb{N}$

$$w_j(x) = \begin{cases} 1, & |x_2| < j, \\ \frac{2j - |x_2|}{j}, & j < |x|_2 < 2j, \\ 0, & \text{else.} \end{cases}$$

The function w_j belongs to $H^1(\Omega)$. Next, we set $v_j = e^{i\alpha x_1} w_j \in H^1_\alpha(\Omega)$. For the gradient of v_j we find

$$|\nabla v_j|^2 = |\nabla w_j|^2 + \alpha^2 |w_j|^2 = |\partial_2 w_j|^2 + \alpha^2 |w_j|^2.$$

Therefore the min-max formula from Lemma 4.1.6 yields

$$\lambda_1(\alpha) \leq \frac{\int_\Omega |\nabla v_j|^2\, dx}{\int_\Omega n^2 |v_j|^2\, dx} \leq \frac{\frac{4\pi}{j} + \alpha^2 \int_\Omega(1 - n^2)|w_j|^2\, dx}{\int_\Omega n^2 |v_j|^2\, dx} + \alpha^2$$

and this quantity is less than α^2 for j large enough. $\qquad\square$

4.1.5   Radiation Condition for Electromagnetic Wave Scattering from Periodic Media

In the last two sections we studied scattering of electromagnetic waves from diffraction gratings using a two-dimensional scattering problem for the Helmholtz equation. Such a simplified model is only valid under special assumptions on the grating and the angle of incidence. Of course, the fully three-dimensional problem of scattering of electromagnetic plane waves from biperiodic dielectric structures governed by Maxwell's equations is of even greater interest for optical applications. In this problem, one meets for instance difficulties caused by a lack of compactness, comparable with the corresponding difficulties occurring in the Maxwell eigenvalue problem in Chapter 2.1.4. We will not go into the details of existence theory here, but only describe the problem setting and set up a radiation condition leading to an expansion of the electromagnetic fields similar to Rayleigh's expansion in the scalar case. This expansion can be coupled to variational formulations for the electromagnetic scattering problems as we did twice before in this chapter. The analytic tools to show existence of solution are then similar to the tools for the Maxwell eigenvalue problem and the scalar scattering problems and we skip this analysis entirely. Analysis of Maxwell's equations in biperiodic media is for instance contained in the papers [3, 21, 1, 2, 20, 11, 47], and the many references therein.

Consider a biperiodic medium described by an electric permittivity $\epsilon : \mathbb{R}^3 \to \mathbb{R}$ and a magnetic permeability $\mu : \mathbb{R}^3 \to \mathbb{R}$ which are biperiodic, that is $\epsilon(x) = \epsilon(x + 2\pi(n_1, n_2, 0)^\top)$, $x \in \mathbb{R}^3$, $n_{1,2} \in \mathbb{Z}$. Further, we assume that there is some constant $h > 0$ such that $\epsilon = \epsilon_0$ and $\mu = \mu_0$ for $|x_3| > h$. The system of governing equations for propagation of monochromatic light described by the electric field E and the magnetic field H in the penetrable structure is

$$\nabla \times E - \mathrm{i}\omega\mu H = 0, \quad \nabla \times H + \mathrm{i}\omega\epsilon E = 0 \quad \text{in } \mathbb{R}^3, \qquad (4.21)$$

where $\omega > 0$ is the angular frequency of the electromagnetic wave (time dependence $\exp(-\mathrm{i}\omega t)$). An incident electromagnetic wave (E^i, H^i) satisfies this system of equations for the background material parameters,

$$\nabla \times E^i - \mathrm{i}\omega\mu_0 H^i = 0, \quad \nabla \times H^i + \mathrm{i}\omega\epsilon_0 E^i = 0 \quad \text{in } \mathbb{R}^3.$$

As for the scalar problem, let us consider a downwards propagating incident plane wave as incident field, with direction $(\sin(\phi_1)\cos(\phi_2), \sin(\phi_1)\sin(\phi_2), \cos(\phi_1))^\top$ where $0 \le \phi_1 < \pi/2$, $0 \le \phi_2 < 2\pi$. It is convenient to define the wave vector

$$k := \omega\sqrt{\epsilon_0\mu_0} \begin{pmatrix} \sin(\phi_1)\cos(\phi_2) \\ \sin(\phi_1)\sin(\phi_2) \\ \cos(\phi_1) \end{pmatrix} =: \begin{pmatrix} \alpha_1 \\ \alpha_2 \\ -\beta \end{pmatrix},$$

since this implies $k \cdot k = \omega^2\mu_+\epsilon_+$. We choose a polarisation $p \in \mathbb{R}^3$ such that $p \cdot k = 0$ and set $q = (k \times p)/(\omega\mu_0)$, and define the incident electromagnetic plane wave by

$$E^i = p e^{ik \cdot x}, \qquad H^i = q e^{ik \cdot x}.$$

These two fields solve Maxwell's equations in free space,

$$\nabla \times E^i = (ik \times p) e^{ik \cdot x} = i\omega\mu_0 q e^{ik \cdot x} = i\omega\mu_0 H^i,$$

$$\nabla \times H^i = ik \times q e^{ik \cdot x} = i\frac{k \times (k \times p)}{\omega\mu_0} e^{ik \cdot x} = i\frac{k(k \cdot p) - p(k \cdot k)}{\omega\mu_0} e^{ik \cdot x} = -i\omega\epsilon_0 E^i.$$

Moreover, E^i, H^i are quasiperiodic in (x_1, x_2) with phase shift (α_1, α_2), that is, $E^i(x + 2\pi(1,0,0)^\top) = \exp(2\pi i\alpha_1) E^i(x)$, $E^i(x + 2\pi(0,1,0)^\top) = \exp(2\pi i\alpha_2) E^i(x)$ for $x \in \mathbb{R}^3$. Set $\alpha = (\alpha_1, \alpha_2)^\top$. Analogous to the scalar case we seek for α-quasiperiodic solutions (E, H) to Maxwell's equations (4.21). In particular, the α-quasiperiodic scattered fields (E^s, H^s) corresponding to incident fields (E^i, H^i) have to satisfy the system

$$\nabla \times E^s - i\omega\mu H^s = i\omega(\mu - \mu_0) H^i,$$

$$\nabla \times H^s + i\omega\epsilon E^s = -i\omega(\epsilon - \epsilon_0) E^i \quad \text{in } \mathbb{R}^3. \quad (4.22)$$

Note that, by assumption on the material parameters, the right-hand side vanishes for $|x_3| > h$.

We are again facing a time-harmonic wave propagation problem in an infinite domain and need to set up a radiation condition for (E^s, H^s) in order to have any chance to obtain uniqueness of solution. We concentrate here on the magnetic field in the half space $\{x_3 > h\}$ and write it as a Fourier series,

$$e^{i(\alpha,0)\cdot x} \cdot H^s(x) = \sum_{j \in \mathbb{Z}^2} H_j(x_3) e^{i(j,0)\cdot x}, \qquad x_3 > h. \quad (4.23)$$

Again, we set $\alpha_j = j + \alpha$, $j \in \mathbb{Z}^2$. Equations (4.22) reduce in $\{x_3 > h\}$ to a second order equation for the magnetic field, $\nabla \times \nabla \times H^s - \omega^2\epsilon_0\mu_0 H^s = 0$ subject to $\nabla \cdot H^s = 0$. The identity $\nabla \times \nabla \times H^s = -\Delta H^s + \nabla(\nabla \cdot H^s)$ implies that $\Delta H^s + \omega^2\epsilon_0\mu_0 H^s = 0$. Hence, separation of variables yields as in Section 4.1.3 a representation of the (upwards radiating) function H^s,

$$H^s(x) = \sum_{j \in \mathbb{Z}^2} H_j e^{i\left((\alpha_j,0)\cdot x + \sqrt{\omega^2\mu_0\epsilon_0 - |\alpha_j|^2}\, x_3\right)}, \qquad x_3 > h.$$

Using this representation of the scattered field, one can derive analogues to the scalar Dirichlet-to-Neumann operator T^+ from Section 4.1.3 and formulate the scattering problem variationally in a bounded domain. This task is beyond the scope of this introductory chapter on scattering theory, but contained in the literature indicated in the beginning of this section.

4.1.6   Scattering from Rough Layers

The scattering problems we discussed up to now describe propagation of time-harmonic waves in bounded or periodic structures. It is also of much practical importance to be able to solve corresponding scattering problems in structures that are unbounded, but not perfectly periodic. This situation arises for instance when a periodic structure is locally perturbed by one or a series of defects. Therefore we give in this section some ideas on how to deal with scattering from *rough unbounded structures*. These structures do not need to possess periodicity, but some smoothness. The word "rough" does hence not refer (as it would be more usual) to the local regularity of the surface but rather to the absence of "global" constraints on the surface variation such as periodicity or decay. Our most important requirements for existence theory are geometric nontrapping conditions on the structure. One central difficulty when dealing with rough structures is their unboundedness which makes compactness arguments usually impossible. To simplify notation we concentrate on the two-dimensional problem, but all results are also valid in $\mathbb{R}^3$.

The rough structures we consider are described by a surface

$$\Gamma = \{(x_1, f(x_1)),\, x_1 \in \mathbb{R}\} \subset \mathbb{R}^2$$

where $f : \mathbb{R} \to \mathbb{R}$, $\delta < f(x_1) < h$, is a differentiable function with bounded and Lipschitz continuous derivative (this space is sometimes denoted as $C_{\mathrm{b}}^{1,1}$) taking values in (δ, h) for $0 < \delta \ll h$. In the domain $\Omega = \{(x_1, x_2)^\top \in \mathbb{R}^2 :\ f(x_1) < x_2\}$ the propagative medium is described by a refractive index $n^2 : \Omega \to \mathbb{R}$ such that $n^2 = 1$ for $x_2 > h - \delta$. Let $\Omega_h := \{(x_1, x_2)^\top \in \mathbb{R}^2 :\ f(x_1) < x_2 < h\}$. We assume that $n^2 \in L^\infty(\Omega)$ satisfies the nontrapping condition

$$\int_{\Omega_h} n^2 \partial_2 \psi\, dx \leq 0 \qquad \text{for all nonnegative } \psi \in C^\infty(\overline{\Omega_h}) \cap L^1(\Omega_h) \text{ with } \psi|_\Gamma = 0.$$
$$(4.24)$$

This condition prevents existence of surface waves (or *trapped modes*) along the rough structure. Note that for a smooth index this condition means that $\partial_2(n^2) \geq 0$. By the way, the condition for existence on surface waves from Section 4.1.4 violates this assumption. Finally, we assume that $n^2 \geq c > 0$ in Ω_h.

The problem we consider is to find a weak solution $u \in H^1_{\mathrm{loc}}(\Omega)$ to a source problem for the Helmholtz equation,

$$\Delta u + k^2 n^2 u = g \quad \text{in } \Omega, \qquad u = 0 \quad \text{on } \Gamma, \qquad (4.25)$$

which describes a time-harmonic wave generated by a source g. This scalar problem arises for the E-mode in a three-dimensional electromagnetic problem, see, e. g., [42].

Of course, we have to impose a radiation condition on u to obtain uniqueness of solution to this problem. Since we use again variational solution theory, one way to proceed is to construct an exterior Dirichlet-to-Neumann operator on the artificial boundary Γ_h, as we did it in Section 4.1.3 for the periodic medium. The boundary Γ_h is now unbounded and Fourier series have to be replaced by the Fourier transform

$$\mathcal{F}\phi(\xi) = (2\pi)^{-1/2} \int_{\mathbb{R}} e^{-i\tilde{x}\cdot\xi}\phi(\tilde{x})\,d\tilde{x}, \quad \xi \in \mathbb{R},$$

which is defined for smooth $\phi \in C_0^\infty(\mathbb{R})$ and extends to a unitary operator on $L^2(\mathbb{R})$ (which we identity with $L^2(\Gamma_h)$). Since a solution u to (4.25) satisfies a Helmholtz equation with constant coefficients in the half-space above Γ_h, one checks that (at least formally)

$$u^\pm(x) = (2\pi)^{-1/2} \int_{\mathbb{R}} e^{i(\pm(x_m - h)\sqrt{k^2 - |\xi|^2} + \tilde{x}\cdot\xi)}\mathcal{F}(u|_{\Gamma_h})(\xi)\,d\xi$$

are two possible solutions which take Dirichlet boundary values u on Γ_h. Naturally, we want to find a time-harmonic wave that propagates upwards (thus, away from the rough structure) and, following the 1D analysis in Section 4.1.1, choose the plus sign in the latter expansion. We remark that this expansion, the *angular spectrum representation*, can be rigorously explained for boundary values in a Sobolev space setting, see [16], but we content us here with a formal computation and take the normal derivative on Γ_h, yielding

$$\partial_2 u|_{\Gamma_h}(\tilde{x}) = i(2\pi)^{-1/2} \int_{\mathbb{R}} \sqrt{k^2 - |\xi|^2}e^{i\tilde{x}\cdot\xi}\mathcal{F}(u|_{\Gamma_h})(\xi)\,d\xi.$$

This formula gives the normal derivative of an upwards radiating function in terms of the Dirichlet values on Γ_h, and hence serves to define a Dirichlet-to-Neumann operator on Γ_h,

$$T : \phi \mapsto i(2\pi)^{-1/2} \int_{\mathbb{R}} \sqrt{k^2 - |\xi|^2}e^{i\tilde{x}\cdot x}\mathcal{F}\phi(\xi)\,d\xi. \tag{4.26}$$

From our study of the periodic scattering problem we know that T needs to be bounded between the Sobolev spaces $H^{\pm 1/2}(\Gamma_h)$ with norm

$$\|\phi\|_{H^s(\Gamma_h)}^2 = \int_{\mathbb{R}^{m-1}} (k^2 + |\xi|^2)^s |\mathcal{F}\phi(\xi)|^2\,d\xi$$

in order to obtain a bounded sesquilinear form for the variational formulation. Achieving this bound is similar as in the periodic case,

$$\|T\phi\|_{H^{-1/2}(\Gamma_h)}^2 = \int_{\mathbb{R}^{m-1}} (k^2 + |\xi|^2)^{-1/2}|\mathcal{F}(T\phi)(\xi)|^2\,d\xi$$

$$= \int_{\mathbb{R}^{m-1}} (k^2 + |\xi|^2)^{1/2}|\mathcal{F}\phi(\xi)|^2\,d\xi = \|T\phi\|_{H^{1/2}(\Gamma_h)}^2.$$

Again, we remark that $H^{1/2}(\Gamma_h)$ is the trace space of $H^1(\Omega_h)$ on Γ_h, and that the trace operator is bounded, $\|v\|_{H^{1/2}(\Gamma_h)} \leq C\|v\|_{H^1(\Omega_h)}$ for $v \in H^1(\Omega_h)$.

The variational formulation corresponding to (4.25) is also similar to the periodic case, the main difference is that the domain Ω_h is now unbounded. Due to the Dirichlet boundary condition on Γ, we seek a solution in $H_0^1(\Omega_h) := \{u \in H^1(\Omega) : u|_\Gamma = 0\}$. We integrate (4.25) against a test function $\bar{v}$, $v \in H_0^1(\Omega_h)$, integrate by parts, and replace $\partial_2 u$ by $T(u)$ in the arising boundary terms on Γ_h,

$$\mathcal{B}_r(u, v) := \int_{\Omega_h} (\nabla u \cdot \nabla \bar{v} - k^2 n^2 u \bar{v}) \, dx - \int_{\Gamma_h} \bar{v} T(u) \, ds = -\int_{\Omega_h} g\bar{v} \, ds \quad (4.27)$$

for all $v \in H_0^1(\Omega_h)$. As Fredholm theory does not apply to this variational problem (Rellich's compactness lemma fails due to the unboundedness of the domain Ω_h), one uses the method of a-priori estimates to prove existence of solution. The a-priori estimate itself can be obtained by a *Rellich identity* and leads to existence of solution through an estimate of the inf-sup constant of the sesquilinear form.

Theorem 4.1.8. *The sesquilinear form $\mathcal{B}_r$ satisfies an inf-sup condition in $H_0^1(\Omega_h)$, that is,*

$$\inf_{u \in H_0^1(\Omega_h)\setminus\{0\}} \sup_{v \in H_0^1(\Omega_h)\setminus\{0\}} \frac{|\mathcal{B}_r(u, v)|}{\|u\|_{H^1(\Omega)}\|v\|_{H^1(\Omega)}} \geq c > 0,$$

together with a corresponding transposed inf-sup condition. The variational problem (4.27) has a unique solution $u \in H_0^1(\Omega_h)$ for all $g \in L^2(\Omega_h)$ with

$$\|u\|_{H^1(\Omega)} \leq \big(1 + h^2/2(1 + k^2)\max(h, 1 + hk)\big)\|g\|_{L^2(\Omega_h)}.$$

We only sketch the proof, notably, we do not prove the *Rellich identity* in full detail.

Proof. First, we establish the bound $\|u\|_{H^1(\Omega_h)} \leq C\|g\|_{L^2(\Omega_h)}$ for a solution to problem (4.27) with an explicit constant C. This is due to the following Rellich identity, which is found by several integrations by parts applied to $\Re \int_{\Omega_h} x_3 \partial_3 \bar{u}(\Delta u - k^2 n^2 u - g) \, dx = 0$,

$$\int_{\Omega_h} (2|\partial_2 u|^2 + k^2 x_2 \partial_2 n^2 |u|^2) \, dx - \Re \int_{\Gamma_h} \bar{u} T(u) \, ds - 2 \int_\Gamma x_2 \nu_2 |\partial_\nu u|^2 \, ds$$

$$= h \int_{\Gamma_h} (|\nabla u|^2 - 2k^2 |u|^2 + 2|\partial_2 u|^2) \, ds - 2\Re \int_{\Omega_h} x_2 \partial_2 \bar{u} g \, dx + \Re \int_{\Omega_h} g\bar{u} \, dx.$$

Here, ν denotes the unit normal field on Γ that points downwards, such that the component ν_2 is negative. The term $\int_{\Omega_h} x_2 \partial_2 n^2 |u|^2 \, dx = -\int_{\Omega_h} n^2 \partial_2 (x_2 |u|^2) \, dx$ is interpreted as a duality pairing between $L^\infty(\Omega_h)$ and $L^1(\Omega_h)$. Note that there appear no boundary terms due to the partial integration since n^2 is constant for $x_2 > h - \delta$ and $|u|^2$ vanishes on Γ. We refer to [16, 37] for a proof of this and similar Rellich identities. Further, it is shown in [16] that the angular spectrum representation (4.26) implies that

$$\int_{\Gamma_h} \left(|\nabla u|^2 - 2k^2|u|^2 + 2|\partial_2 u|^2 \right) ds \le 2k \Im \left(\int_{\Omega_h} g\bar{u} \, dx \right).$$

We recall that Poincaré's inequality $\|u\|_{L^2(\Omega_h)} \le h/\sqrt{2} \, \|\partial_2 u\|_{L^2(\Omega_h)}$ holds for $u \in H_0^1(\Omega_h)$ because the domain Ω_h is bounded in one direction, and again refer to [16] for a proof. Finally, we note that $-\Re \int_{\Gamma_h} \bar{\phi} T(\phi) \, ds \ge 0$ for any $\phi \in H^{1/2}(\Gamma_h)$. Hence, the Rellich identity, the nontrapping condition (4.24) and the Poincaré inequality imply

$$\begin{aligned}
4/h^2 \, \|u\|_{L^2(\Omega_h)}^2 &\le 2\|\partial_2 u\|_{L^2(\Omega_h)}^2 \\
&\le \|g\|_{L^2(\Omega_h)} \left(2h\|\partial_2 u\|_{L^2(\Omega_h)} + (1 + 2hk)\|u\|_{L^2(\Omega_h)} \right) \\
&\le \max(2h, 1 + 2hk)\|g\|_{L^2(\Omega_h)}\|u\|_{H^1(\Omega_h)}.
\end{aligned}$$

Recall that $n^2 \ge c$ in Ω_h and that, necessarily, $c \le 1$. From the real part of the variational formulation with $v = u$ we obtain that $\|\nabla u\|_{L^2(\Omega_h)}^2 - k^2\|u\|_{L^2(\Omega_h)}^2 \le \|g\|_{L^2(\Omega_h)}\|u\|_{L^2(\Omega_h)}$. In consequence,

$$\begin{aligned}
\|u\|_{H^1(\Omega_h)}^2 &\le (1 + k^2)\|u\|_{L^2(\Omega_h)}^2 + \|g\|_{L^2(\Omega_h)}\|u\|_{L^2(\Omega_h)} \\
&\le (1 + k^2)h^2/4 \max(2h, 1 + 2hk)\|g\|_{L^2(\Omega_h)}\|u\|_{H^1(\Omega_h)} + \|g\|_{L^2(\Omega_h)}\|u\|_{L^2(\Omega_h)} \\
&\le (1 + h^2/2(1 + k^2) \max(h, 1 + hk))\|g\|_{L^2(\Omega_h)}\|u\|_{H^1(\Omega_h)}.
\end{aligned}$$

This is an a-priori estimate in $H_0^1(\Omega_h)$ for a solution to the variational formulation with right-hand side in $L^2(\Omega_h)$. To establish the inf-sup condition claimed in the theorem, we need to extend this estimate to an arbitrary anti-linear form $G \in H_0^1(\Omega_h)^*$ in the dual of $H_0^1(\Omega_h)$ as right hand side. One uses that the difference between a solution to the coercive problem to find $u_0 \in H_0^1(\Omega_h)$ with $\int_{\Omega_h} (\nabla u_0 \cdot \nabla \bar{v} + u_0 \bar{v}) \, dx = G(v)$ for all $v \in H_0^1(\Omega_h)$, and the variational formulation of the Helmholtz equation (4.27) with right-hand side G satisfies again (4.27) with right hand side $k^2(n^2 + 1)u_0 \in L^2(\Omega_h)$. Hence, for this difference, the above a-priori estimate applies. Since an estimate for u_0 is easily obtained by coercivity, the triangle inequality shows that

$$\|u\|_{H^1(\Omega_h)} \le \underbrace{1 + 2k^2\left(1 + h^2/2(1 + k^2) \max(h, 1 + hk)\right)}_{=:\gamma} \|G\|_{H_0^1(\Omega_h)^*}.$$

Therefore

$$\sup_{v \in H_0^1(\Omega_h)\setminus\{0\}} \frac{|\mathcal{B}_r(u, v)|}{\|v\|_{H^1(\Omega_h)}} \ge \gamma^{-1}\|u\|_{H^1(\Omega_h)} > 0.$$

From symmetry properties of the sesquilinear form $\mathcal{B}_r$ shown in [16, Corollary 3.3 & Lemma 4.2] it follows that this directly implies a transposed inf-sup condition. We can conclude by generalized Lax-Milgram theory (see [9] or [41, Theorem 2.22]) that the variational problem (4.27) is uniquely solvable. $\qquad \square$

4.2   Time-Harmonic Inverse Wave Scattering

In an inverse scattering problem, the task is to determine features of a scattering object from partial knowledge of waves scattered from this object. Such features might include geometric information or physical properties such as values of certain material parameters. As a model problem, we study in the following the determination of the shape of a bounded scatterer from far field measurements and of a periodic structure from near field measurements. We studied the corresponding direct time-harmonic scattering problems in the first part of this chapter. Notably, a thorough understanding of the direct problem is usually inevitable for a reasonable analysis of the inverse problem. The method we employ for both problems under investigation is the *Factorisation method*, it is a member of the class of so-called *sampling methods* for inverse problems. These methods, introduced in [18, 32], recently gained some popularity since they provide geometric information in short time compared to "classical" optimisation-type methods that try to characterise the physical properties of the scattering object more completely. We refer to [33, 14] for more details on sampling methods. Our material for inverse scattering from bounded inhomogeneous media follows [33, Chapter 4] and [37]; the material on the periodic scattering problem is from [38, 36] and [5].

Generally speaking, inverse scattering problems have a very much different character than direct wave propagation problems: They are nonlinear and ill-posed. According to Hadamard's definition from [27], a problem (that is, an operator equation between Banach spaces) is called well-posed if there exists a unique solution for any right-hand side, and if this solution depends continuously on the data. If a problem does not meet these requirements, we call it ill-posed. Inverse scattering problems belong to the latter class of problems since the operator mapping the scatterer to the measurements is discontinuous in any reasonable topology – more insight on this behaviour is given in [19]. An ill-posed problem needs a special treatment called regularisation to obtain a stable solution, and regularisation theory of inverse and ill-posed problems is by now a well-established subject. Introductions to this theory are provided in, e. g., [31, 45, 24].

4.2.1   The Factorisation Method for Bounded Penetrable Objects

In Section 4.1.2 we considered scattering of a plane wave $u^i(\cdot, \theta) = \exp(ik\,\theta \cdot x)$ of direction $\theta \in \mathbb{S}$ from an inhomogeneous medium described by a refractive index $n^2 : \mathbb{R}^3 \to \mathbb{C}$. We have seen that the corresponding scattered field $u^s(\cdot, \theta)$ behaves far away from the scatterer like a spherical wave with amplitude $u^s_\infty(\cdot, \theta)$. The inverse problem we consider now is to find the (bounded) support $\overline{D}$ of the contrast $q = n^2 - 1$ when given the far field pattern $u_\infty(\hat{x}, \theta)$ for all angles of measurement $\hat{x} \in \mathbb{S}$ and all incident directions $\theta \in \mathbb{S}$. We construct an explicit formula for reconstructing (more precisely, for imaging) D from this data, by a technique called the Factorisation method. To apply this technique, we will need to strengthen our assumptions on the contrast q from Section 4.1.2. Again, we concentrate here on a three-dimensional setting, even if all theory also holds in two dimensions.

From now on for the rest of this chapter we assume that q has the following properties.

Assumption 4.2.1. The contrast $q \in L^\infty(\mathbb{R}^3)$ satisfies $\operatorname{supp}(q) = \overline{D}$ where D is a Lipschitz domain in $\mathbb{R}^3$ with connected complement. Moreover, we suppose that $\Re(q) \geq c$ for some $c > 0$ and $\Im(q) \geq 0$ almost everywhere in D.

These assumptions on the contrast are not the most general ones but easy to treat in the analysis later on. See [33, 37] for assumptions that allow, e. g., for q that vanishes on the boundary or on isolated points in the interior of D. However, it is currently an open problem whether or not our assumption that the sign of q does not change is necessary for the method to work.

Central to our method to solve the inverse scattering problem is the far field operator,

$$F : L^2(\mathbb{S}) \to L^2(\mathbb{S}), \qquad g \mapsto \int_{\mathbb{S}} g(\theta)\, u_\infty(\cdot, \theta)\mathrm{d}s(\theta)\,.$$

This is an integral operator with continuous (even analytic) kernel $u_\infty(\cdot, \cdot)$, see, e.g, [19]. Therefore, F is a compact operator on $L^2(\mathbb{S})$. By linearity of the scattering problem, F maps a density g to the far field corresponding to the incident Herglotz wave function

$$v_g(x) = \int_{\mathbb{S}} g(\theta)u^i(\cdot, \theta)\mathrm{d}s(\theta) = \int_{\mathbb{S}} g(\theta)e^{ik\,\theta\cdot x}\mathrm{d}s(\theta)\,, \qquad x \in \mathbb{R}^3.$$

Obviously, if we know $\{u_\infty(\hat{x}, \theta) : \hat{x}, \theta \in \mathbb{S}\}$ for all directions $\hat{x}, \theta \in \mathbb{S}$, then we also know F. Therefore we reformulate the inverse scattering problem as follows: Given F, determine the support D of the contrast q!

Colton and Kirsch [18] first showed that factorisations of the far field operator can be particularly useful for reconstruction of D. Their observation was based on the Herglotz operator,

$$H : L^2(\mathbb{S}) \to L^2(D), \qquad g \mapsto \int_{\mathbb{S}} g(\theta)e^{ik\,\theta\cdot x}\mathrm{d}s(\theta)\,, \qquad x \in D,$$

which is the restriction of the Herglotz wave function v_g to D. The Cauchy-Schwartz inequality implies that H is a bounded linear operator. The far field pattern v_∞ of the scatterer wave for incident field v_g is $Fg = v_\infty$. We define the *solution operator* $G : L^2(D) \to L^2(\mathbb{S})$ by $f \mapsto v_\infty$, where v_∞ is the far field pattern of v, solution to the scattering problem (4.10), i. e., $\Delta v + k^2(1 + q)v = -k^2 q f$ in $\mathbb{R}^3$. Then $F = GH$ holds by construction of the far field operator.

The range of the solution operator G can be used to characterise the scatterer's support D, as the following lemma shows.

Lemma 4.2.2. *The far field $\Phi_\infty(\cdot, z)$ of a point source $\Phi(\cdot, z)$ at $z \in \mathbb{R}^3$ belongs to the range of G if and only if $z \in D$.*

Proof. We first show that the far field of a point source at $z \in \mathbb{R}^3 \setminus D$ does not belong to the range of G: Assume, on the contrary, that v is a radiating solution of (4.10) for some $f \in L^2(D)$ with $v_\infty = \Phi_\infty(\cdot, z)$, then the far field pattern of the radiating solution $w = v - \Phi(\cdot, z)$ vanishes. Lemma 4.1.2 implies that $v = \Phi(\cdot, y)$ in $\mathbb{R}^3 \setminus (D \cup \{z\})$, however, $\Phi(\cdot, y)$ has a first-order singularity at y but $v \in H^2_{\text{loc}}(\mathbb{R}^3)$, which is a contradiction.

If $z \in D$, then we choose a small open ball B around z such that $\overline{B} \subset D$, and a smooth function $\chi \in C_0^\infty(B)$ with $\chi = 1$ in a neighbourhood of z. Let us extend χ by zero to all of $\mathbb{R}^3$. Then

$$
\Delta \left((1 - \chi)\Phi(\cdot, y)\right) + k^2 n^2 (1 - \chi)\Phi(\cdot, y)
$$
$$
= -k^2 q \left(-q^{-1}\Delta\chi\Phi(\cdot, y) - 2q^{-1}\nabla\chi \cdot \nabla\Phi(\cdot, y) + k^2(1 - \chi)\Phi(\cdot, y)\right) =: -k^2 q f,
$$

and f belongs to $L^2(D)$ because $\Re(q) \geq c > 0$. Finally, $((1 - \chi)\Phi(\cdot, y))_\infty = \Phi_\infty(\cdot, y)$ shows that $G(f) = v_\infty$. $\qquad\square$

Unfortunately, the solution operator G is unknown to us as D is unknown, and we only know the measurement operator $F = GH$. Since the Herglotz operator H is compact, the range of F is in general different from the range of G (but we will see below that there is a link between them). Nevertheless, Colton and Kirsch [18] proposed in 1996 to consider the far field equation $Fg_z = \Phi_\infty(\cdot, z)$ in $L^2(\mathbb{S})$ and to check for each z whether this linear integral equation has a solution. However, F being a compact operator, this linear problem is ill-posed, even worse, in general there will be no solution to the far field equation. Hence, one applies a regularisation strategy, and solves $\epsilon g_z^\epsilon + F^* F g_z^\epsilon = F^* \Phi_\infty(\cdot, z)$ for a particularly chosen regularisation parameter $\epsilon > 0$. This is the Tikhonov regularisation applied to the far field equation, see, e. g., [14] for details. The Linear Sampling Method then consists in plotting $1/\|\phi_z^\epsilon\|_{L^2(\mathbb{S})}$ as a function of z. The idea is that for $z \notin \overline{D}$, $\Phi_\infty(\cdot, z)$ cannot belong to $\text{Rg}(G) \supset \text{Rg}(F)$, thus, $\|g_z^\epsilon\|$ is expected to be large. On the other hand, for $z \in D$, $\Phi_\infty(\cdot, z) \in \text{Rg}(G)$ and one might hope that therefore the regularized solution to the far field equation also has much smaller norm than outside D. Consequently, plotting $z \mapsto 1/\|\phi_z^\epsilon\|_{L^2(\mathbb{S})}$ on a grid should result in small values outside D and considerably larger ones inside D. This vague statement can be formulated in a mathematically more satisfactory way, see [14], and in numerical experiments one observes that the method works very well indeed. But it took a long time until a rigorous explanation was found why (a slightly different version of) the Linear Sampling method does so [7, 4]. Let us remark that the crucial ingredient of this explanation is a strong link between the Linear Sampling method and the so-called Factorisation method. This method, which we investigate in the following, gives a direct description of the range of G in terms of F, which characterises the obstacle D in turn by Lemma 4.2.2. The following refinement of the far field operator's factorisation will be very helpful — and also explains the method's name.

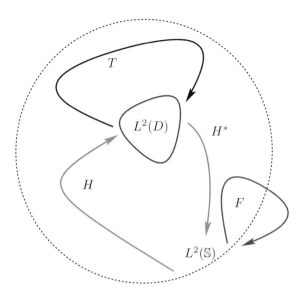

Figure 4.5: Factorisation of the far field operator $F = H^*TH$. The Herglotz operator H maps a density on the unit sphere to the restriction of the Herglotz wave function with that density to the obstacle. The middle operator T involves the solution of a scattering problem, and H^* gives the far field pattern of the solution.

Theorem 4.2.3. *The far field operator can be factored as*

$$F = H^*TH,$$

where $T : L^2(D) \to L^2(D)$ is defined by $Tf = k^2q\,(f + v|_D)$, and $v \in H^1_{\mathrm{loc}}(\mathbb{R}^3)$ is the radiating solution of

$$\Delta v + k^2(1+q)\,v = -k^2 qf \qquad in \ \mathbb{R}^3. \tag{4.28}$$

The far field operator's factorisation is illustrated in Figure 4.5. We remark that in two dimensions a similar factorisation holds but involves an additional constant, see [36, 14].

Proof. The adjoint of H with respect to the inner product of $L^2(D)$ is easily computed to be

$$H^* : L^2(D) \to L^2(\mathbb{S}), \quad f \mapsto \int_D f(x)e^{-ik\,\theta\cdot x}\,dx, \quad \theta \in \mathbb{S}.$$

In Theorem 4.1.4 we showed that the Helmholtz equation (4.28) together with the Sommerfeld radiation condition has a unique solution $v \in H^1_{\mathrm{loc}}(\mathbb{R}^3)$ given by the volume potential

$$v = k^2 V\,(q(f + v|_D))\,. \tag{4.29}$$

The adjoint H^* maps $f \in L^2(D)$ to the far field pattern of the volume potential $V(f)$, since the far field pattern of $x \mapsto \Phi(x,y)$ is $\exp(-ik\,\hat{x} \cdot y)$ (see (4.9)). In consequence, $H^*f = (Vf)_\infty$. Now, plug in Tf for f in the last formula, to see that

$$H^*Tf = k^2 \left(V\big(q(f + v|_D)\big) \right)_\infty .$$

Comparison of right-hand side of this equation with (4.29) shows that H^*Tf is the far field pattern of the radiating solution to (4.28): $H^*T = G$. This proves the claimed factorisation for F. ∎

Imaging by Factorisation

The Herglotz wave operator H, the middle operator T and also the solution operator G have certain special properties which we will exploit for construction of the Factorisation method. We gather these properties in the following lemma.

Lemma 4.2.4. *Assume that the contrast q satisfies Assumption 4.2.1. Then the following statements hold.*

(a) *The Herglotz operator $H : L^2(\mathbb{S}^2) \to L^2(D)$ is compact and injective.*
(b) *The middle operator $T : L^2(D) \to L^2(D)$ is injective and has a natural splitting $T = T_1 + T_0$ into a sum of a coercive operator T_0, $T_0 f := k^2 q f$, and a compact operator T_1, $T_1 f = k^2 q\, v|_D$, where v is again the radiating solution of (4.28). Moreover, T is an isomorphism and*

$$\Im \left(\int_D \overline{f}\, Tf \, dx \right) \geq 0 \qquad \text{for } f \in L^2(D). \tag{4.30}$$

Proof. (a) The kernel of H is a smooth function, hence H is a smoothing operator, and therefore compact by Rellich's embedding theorem. If $Hg = 0$ in D, then the entire solution to the Helmholtz equation

$$x \mapsto \int_{\mathbb{S}} g(\theta) e^{ik\,\theta \cdot x} ds(\theta)$$

vanishes in D. The unique continuation property directly implies that Hg vanishes entirely in $\mathbb{R}^3$ and the Jacobi–Anger expansion [19, Chapter 2] shows that g vanishes.

(b) T_1 is a compact operator due to Rellich's compact embedding theorem: The solution v to (4.28) belongs to $H^2(B)$ for all open balls $B \supset D$, and compactness of the embedding $H^2(B) \hookrightarrow L^2(B)$ implies compactness of $L^2(D) \ni f \mapsto v|_D \in L^2(D)$. Further, multiplication by $k^2 q$ is a bounded operation on $L^2(D)$. Therefore T_1 is compact. Concerning T_0, Assumption 4.2.1 implies

$$\Re \left(\int_D \overline{f}\, T_0 f \, dx \right) = k^2 \int_D \Re(q)|f|^2 \, dx \geq ck^2 \|f\|_{L^2(D)}^2,$$

that is, T_0 is coercive on $L^2(D)$.

Next we prove that T is injective. If $Tf = 0$, then $f = -v$ in D. Since v is a radiating solution of $\Delta v + k^2(1+q)\, v = -k^2 qf$, we have $\Delta v + k^2(1+q)\, v = k^2 qv$ in $\mathbb{R}^3$. Thus, v is an entire radiating solution of $\Delta v + k^2 v = 0$ in all of $\mathbb{R}^3$, but this implies that v vanishes. Consequently, T in an injective Fredholm operator of index zero, and hence an isomorphism.

Finally, we show (4.30). For $f \in L^2(D)$,

$$\int_D \overline{f}\, Tf\, dx = k^2 \int_D q w \overline{f}\, dx = k^2 \int_D q \left(|w|^2 - w\overline{v}\right)\, dx, \tag{4.31}$$

with $w := f + v|_D$. The definition of v in (4.28) yields $\Delta v + k^2 v = -k^2 qw$. We plug this relation into (4.31) and find by Green's first identity that

$$\int_D \overline{f}\, Tf\, dx = k^2 \int_D q|w|^2\, dx + \int_D \left(\Delta v + k^2 v\right) \overline{v}\, dx$$

$$= k^2 \int_D q|w|^2\, dx + \int_D \left(k^2|v|^2 - |\nabla v|^2\right)\, dx + \int_{\partial D} \overline{v}\, \partial_\nu v\, ds\,.$$

In the last integral, ν denotes the exterior unit normal to D. Since $\Im(q) \geq 0$ by assumption, we only need to show that $\Im\left(\int_{\partial D} \overline{v}\, \partial v / \partial \nu\, ds\right) \geq 0$. However, assuming that $\Im\left(\int_{\partial D} \overline{v}\, \partial v / \partial \nu\, ds\right) < 0$ Lemma 4.1.2(c) yields that v vanishes entirely, which is a contradiction. $\qquad\square$

The preparations made in the last lemma combined with an abstract result on range identities, see Theorem 4.2.5, directly characterises D in terms of F. As a prerequisite, we introduce real and imaginary part of a bounded linear operator. Let $\mathsf{X} \subset \mathsf{U} \subset \mathsf{X}^*$ be a Gelfand triple, that is, U is a Hilbert space, X is a reflexive Banach space with dual X^* for the inner product of U, and the embeddings are injective and dense. Then real and imaginary part of a bounded operator $T : \mathsf{X}^* \to \mathsf{X}$ are defined in accordance with the corresponding definition for complex numbers,

$$\Re(T) := \frac{1}{2}(T + T^*), \qquad \Im(T) := \frac{1}{2\mathrm{i}}(T - T^*).$$

Theorem 4.2.5. *Let $\mathsf{X} \subset \mathsf{U} \subset \mathsf{X}^*$ be a Gelfand triple with Hilbert space U and reflexive Banach space X. Furthermore, let V be a second Hilbert space and $F : \mathsf{V} \to \mathsf{V}$, $H : \mathsf{V} \to \mathsf{X}$ and $T : \mathsf{X} \to \mathsf{X}^*$ be linear and bounded operators with*

$$F = H^* T H.$$

We make the following assumptions:

(a) *H is compact and injective.*

(b) *$\Re(T)$ has the form $\Re(T) = T_0 + T_1$ with a coercive operator T_0 and a compact operator $T_1 : \mathsf{X} \to \mathsf{X}^*$.*

(c) *$\Im(T)$ is nonnegative on X: $\langle \Im(T)\phi, \phi \rangle \geq 0$ for all $\phi \in \mathsf{X}$.*

(d) *T is injective.*

Then $F_\sharp := |\Re(F)| + \Im(F)$ is positive definite and the ranges of $H^* : X^* \to V$ and $F_\sharp^{1/2} : V \to V$ coincide.

For a proof of this result we refer to [37].

Theorem 4.2.6. *Let $(\lambda_j, \psi_j)_{j \in \mathbb{N}}$ be an orthonormal basis of the selfadjoint compact operator $F_\sharp := |\Re(F)| + \Im(F)$. A point $y \in \mathbb{R}^3$ belongs to D if and only if*

$$\sum_{j=1}^\infty \frac{\left| \langle \Phi_\infty(\cdot, y), \psi_j \rangle_{L^2(\mathbb{S})} \right|^2}{\lambda_j} < \infty. \tag{4.32}$$

Proof. The claim follows directly from an application of Theorem 4.2.5 to the factorisation $4\pi F = H^* T H$ with $V = L^2(\mathbb{S})$, $X = X^* = L^2(D)$. All the assumptions of Theorem 4.2.5 have been checked in Lemma 4.2.4. Note that (4.30) implies that $\Im(T)$ is positive semidefinite, since $\Im(\int_D \bar{f} T f \, dx) = \int_D \bar{f} \Im(T) f \, dx$. Therefore, Theorem 4.2.5 implies that $\mathrm{Rg}(F_\sharp^{1/2}) = \mathrm{Rg}(H^*)$. Using Theorem 4.2.3, $\mathrm{Rg}(H^*) = \mathrm{Rg}(T^{-1}G) = \mathrm{Rg}(G)$. Since (λ_j, ψ_j) is an eigensystem of $F_\sharp$ we observe that $(\lambda_j^{1/2}, \psi_j)$ is an eigensystem of $F_\sharp^{1/2}$ and Picard's criterion [31] implies that

$$f \in L^2(\mathbb{S}) \text{ belongs to } \mathrm{Rg}(G) \quad \Longleftrightarrow \quad \sum_{j=1}^\infty \frac{|\langle f, \psi_j \rangle|^2}{\lambda_j} < \infty.$$

Lemma 4.2.2 finally yields the claim of the theorem. $\square$

Recall from (4.9) that the far field $\Phi_\infty(\cdot, y)$ is given by $\hat{x} \mapsto \exp(-\mathrm{i}k\,\hat{x}\cdot y)$. In a numerical implementation of the Factorisation method, one picks a discrete set of grid points in a certain test domain and plots the inverse of the value of the series in (4.32) on this grid. Of course, we cannot numerically compute the entire series but only a finite approximation afflicted with certain errors. Nevertheless, one might hope that plotting the truncated series as a function of y leads to small values at points z outside the support D of the contrast q and to large values at points inside D. But it is crucial to note that ill-posedness certainly afflicts the imaging process, because we divide by small numbers λ_j. If we only know approximations λ_j^δ with $|\lambda_j^\delta - \lambda_j| \leq \delta$, then the error in the term $|\langle \Phi_\infty(\cdot, z), \psi_j \rangle|^2 / \lambda_j^\delta$ is in general expected to be much larger than δ. Even worse, also the singular vectors will be afflicted with a certain numerical error, and we only know ψ_j^δ with $\|\psi_j - \psi_j^\delta\| < \delta$. Hence, it seems necessary to regularise the series (4.32). Several methods are available: truncation of the series at some index corresponding to the regularisation parameter [35], comparison techniques [28] and Tikhonov regularisation [17], which means to plot a noisy version of

$$z \mapsto \left(\sum_{j=1}^\infty \frac{\lambda_j}{(\lambda_j + \epsilon)^2} \left| \langle \Phi_\infty(\cdot, z), \psi_j \rangle \right|^2 \right)^{-1} \tag{4.33}$$

for an adequately chosen regularisation parameter. For the following numerical examples, we use this function to obtain an image of the scatterer. The eigensystem (λ_j, ψ_j) is then replaced by numerical approximations $(\lambda_j^\delta, \psi_j^\delta)$. The synthetic data is additionally perturbed by random noise (a random matrix). Each measurement is perturbed by a uniformly distributed random variable and the norm of the artificially added matrix is scaled to a certain noise level.

In Figure 4.6 and 4.7 we use a numerical approximation of the far field operator of dimension 32. For all plots, the series is truncated at $N = 20$ and ϵ is chosen to be twice the noise level of the data. When no artificial noise is added, the noise level is found through the singular values of the measurement matrix, in the same way as it is described, e. g., in [8]. The artificial noise is in general different for all plots. Figure 4.6 presents reconstructions of a nonconvex but connected inhomogeneity, whereas Figure 4.6 shows reconstructions of a disconnected inhomogeneity consisting of two convex objects. The wave numbers 3π and 2π are both times about the size of the object.

Actually, the exponential decay of the singular values implies that only a very restricted number of the λ_j can be obtained in a reliable way. In our examples this number is round about 10 (depending essentially on the wave number and the size of the object), even if we do not add artificial noise to the data. So, the *data subspace* containing those singular vectors that possess reasonable accuracy of, say, one percent relative error is always rather low-dimensional compared to the *noise subspace*, which contains all the remaining singular vectors. In consequence, a lot of the terms $\left|\langle \Phi_\infty(\cdot, z), \psi_j^\delta \rangle\right|$ will have *nothing* to do with the corresponding exact term (i. e., for $\delta = 0$). Nevertheless, there is a reason to use these terms for imaging: Since singular vectors are orthogonal to each other, the noise subspace is always orthogonal to the data subspace. The paper [8] explains how this substantial information can be exploited for imaging.

4.2.2 The Factorisation Method for Diffraction Gratings

In this section we use the Factorisation method to determine a periodic dielectric structure mounted on a perfectly conducting plate. This problem arises for instance in nondestructive testing of periodic diffraction gratings, where one uses incident monochromatic light (a laser beam) to illuminate a periodic structure under different angles and measures the corresponding reflected waves. From these measurements one aims to determine the periodic structure. An important class of diffraction gratings are constant in one lateral direction and periodic in the orthogonal lateral direction. For these structures, the direct and inverse electromagnetic scattering problems decouple into two scalar problems when the wave vector of the incident field is orthogonal to the invariance axis of the incident fields. In this section, we only treat one of the two reduced scalar problems corresponding to E-polarisation. Details on the physical background and practical applications of direct and inverse scattering problems for diffraction gratings can be found, e. g., in [26].

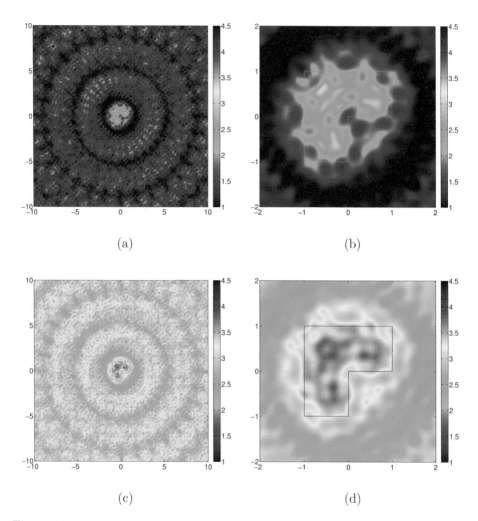

(a)   (b)

(c)   (d)

Figure 4.6: Inverse scattering for the L-shaped inhomogeneous medium from Figure 4.2. The wave number equals 3π, the contrast inside the obstacle equals 10, and we use 32 uniformly distributed incident plane waves and measurement directions. The boundary of the obstacle is indicated by a thin line. The four plots show regularised truncated indicator function of the factorisation method, see (4.33). (a) Plot on $[-10, 10]^2$, no artificial noise (b) Plot on $[-2, 2]^2$, no artificial noise (c) Plot on $[-10, 10]^2$, relative noise level of five percent (d) Plot on $[-2, 2]^2$, relative noise level of five percent.

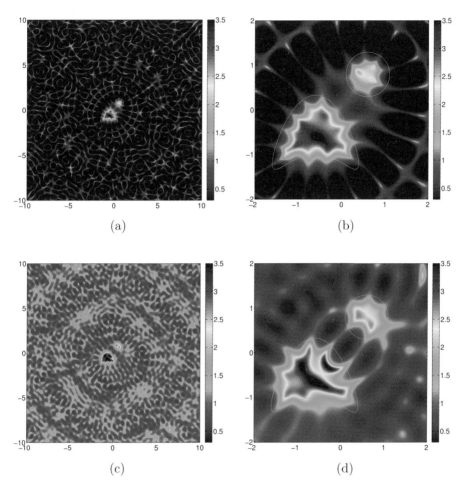

Figure 4.7: Inverse scattering for a disconnected inhomogeneous medium consisting of a kite and a circle. The wave number k equals 2π, the contrast inside the obstacle equals 0.2, and we use 32 uniformly distributed incident plane waves and measurement directions. The boundary of the obstacle is indicated by a thin line. The four plots show regularised truncated indicator function of the factorisation method, see (4.33). (a) Plot on $[-10, 10]^2$, no artificial noise (b) Plot on $[-2, 2]^2$, no artificial noise (c) Plot on $[-10, 10]^2$, relative noise level of five percent (d) Plot on $[-2, 2]^2$, relative noise level of five percent.

For imaging of periodic structures there are principally the same techniques available as for bounded objects, but their analysis is sometimes more involved. To give only a few references, the paper [22] studies an optimisation approach for grating profile reconstruction. In [13], the authors study a continuation technique called the Kirsch–Kress method for grating profile reconstructions. Further, in [6, 5] the authors study a Factorisation method for determination of a periodic surface with Dirichlet boundary condition. Some of the ideas in the material below originate in these works. The problem we investigate is close to reference [38], where a similar periodic inverse medium problem for a dielectric structure is investigated in full space, that is, without a conducting plate below the structure. Generally speaking, the method we describe is an instance of near field tomography, since we require data measured a finite distance (in practice less than one wavelength) away from the periodic structure. In contrast, the data we required in the last section has been far field data which is in practice measured several wavelengths away from the object. More on near field tomography (in the Born regime) can be found in [15], where also lots of different associated measurement modalities are explained. Inclusion of such physical properties of the measurement process is however beyond our scope here.

Our setting for the periodic inverse scattering is a half-space variant to the periodic direct scattering problem in Section 4.1.3. Since we want to model the metallic plate supporting the dielectric material, the problem is placed in

$$\mathbb{R}^2_+ = \{(x_1, x_2)^\top, x_2 > 0\}.$$

We consider a periodic index of refraction $n^2 \in L^\infty(\mathbb{R}^2_+)$, $\Re(n^2) \geq c > 0$, $\Im(n^2) \geq 0$. Again $n^2(x_1 + 2\pi, x_2) = n^2(x_1, x_2)$ for $(x_1, x_2)^\top \in \mathbb{R}^2$ and $n^2 = 1$ for $x_2 > h - \delta$, $0 < \delta \ll h$. The wave number is again denoted by k and as in Section 4.1.3 we assume that $k^2 \neq \alpha_j^2$ for any $j \in \mathbb{Z}$. The incident fields we use to obtain data for the Factorisation method are α-quasiperiodic plane waves, where $\alpha \in [0, 1)$ is fixed throughout the section. The total field u solves $\Delta u + k^2 n^2 u = 0$ in $\mathbb{R}^2_+$ and $u = 0$ on $\Gamma_0 := \{(x_1, x_2)^\top : -\pi < x_1 < \pi, x_2 = 0\}$, and the scattered field $u^s = u - u^i$ satisfies the Rayleigh expansion condition (4.13). The differential equation for u^s is hence $\Delta u^s + k^2 n^2 u^s = -k^2 q u^i$, subject to $u^s = -u^i$ on Γ_0. Actually, it is convenient here (and suffices for later purposes) to merely consider incident fields of the form

$$u_j^i = e^{i(\alpha_j x_1 - \beta_j x_2)} - e^{i(\alpha_j x_1 + \beta_j x_2)}, \quad j \in \mathbb{Z}, \tag{4.34}$$

which vanish on Γ_0. Recall that $\alpha_j = j + \alpha$ and $\beta_j := (k^2 - \alpha_j^2)^{1/2}$ for $j \in \mathbb{Z}$. Since we assume that $\alpha_j^2 \neq k^2$ for all $j \in \mathbb{Z}$, we have $\beta_j \neq 0$. For the given incident fields, $u^s = -u_j^i = 0$ on Γ_0.

The incident wave u_j^i is a linear combination of $\exp(i(\alpha_j x_1 - \beta_j x_2))$, a downward propagating field, and $\exp(i(\alpha_j x_1 + \beta_j x_2))$, which is an upward propagating field. The upward propagating part has no physical meaning, but from an abstract mathematical viewpoint it poses no difficulty: The scattered field corresponding to such an incident field equals minus the incident field. This special choice for the incident fields u_j^i has been proposed and discussed in [5]; remarkably, concerning the analysis of the Factorisation method, these fields offer some crucial advantages over, e. g., simple incident plane waves $\exp(\mathrm{i}(\alpha_j x_1 - \beta_j x_2))$.

As in Section 4.1.3 we seek for a variational solution in the domain $\Omega_h := (-\pi, \pi) \times (0, h)$ and denote $\Gamma_h = (-\pi, \pi) \times \{h\}$. Analogously to (4.15) we define the quasiperiodic Sobolev space $H_\alpha^1(\Omega_h)$ and set $H_{\alpha,0}^1(\Omega_h) = \{u \in H_\alpha^1(\Omega_h),\, u = 0 \text{ on } \Gamma_0\}$. The variational formulation for the scattered field $u_j^s \in H_{\alpha,0}^1(\Omega_h)$ solving $\Delta u_j^s + k^2(1 + q)u_j^s = -k^2 q u_j^i$, subject to the Rayleigh expansion condition in the form $\partial_2 u_j^s = T^+(u_j^s)$ on Γ_h, is

$$\int_{\Omega_h} (\nabla u_j^s \cdot \nabla \overline{v} - k^2 n^2 u_j^s \overline{v})\, ds - \int_{\Gamma_h} \overline{v}\, T^+(u_j^s)\, ds = k^2 \int_{\Omega_h} q u_j^i \overline{v}\, dx \qquad (4.35)$$

for all $v \in H_{\alpha,0}^1(\Omega_h)$. A more general problem is to find $u_j^s \in H_{\alpha,0}^1(\Omega_h)$ such that

$$\int_{\Omega_h} (\nabla u_j^s \cdot \nabla \overline{v} - k^2 n^2 u_j^s \overline{v})\, ds - \int_{\Gamma_h} \overline{v}\, T^+(u_j^s)\, ds = k^2 \int_{\Omega_h} q f \overline{v}\, dx \qquad (4.36)$$

for all $v \in H_{\alpha,0}^1(\Omega_h)$, where $f \in L^2(\Omega_h)$ is a given source function. Existence theory for this problem can be derived in the same way as we did it in Section 4.1.3; especially, existence and uniqueness hold if $\Im(n^2) > 0$ in some open subset of Ω_h, or if k does not belong to the (countably many) exceptional wave numbers. In the sequel, we will simply *assume* that (4.35) is uniquely solvable for any $j \in \mathbb{Z}$. Then we can define a *solution operator* $G : L^2(\Omega_h) \to \ell^2$ which maps f to the Rayleigh sequence of $(\hat{u}_j)_{j \in \mathbb{Z}}$ of $u \in H_{\alpha,0}^1(\Omega_h)$, solution to (4.36).

If one combines several incident fields, the resulting scattered field is the corresponding linear combination of the scattered fields. We achieve such linear combination using sequences $(a_j)_{n \in \mathbb{Z}} \in \ell^2$ and define the corresponding operator by

$$H(a_j) = \sum_{j \in \mathbb{Z}} \frac{a_j}{\beta_j w_j}\, u_j^i\big|_{\Omega_h}, \qquad \text{where } w_j := \begin{cases} 1, & k^2 > \alpha_j^2, \\ \exp(-\mathrm{i}\beta_j h), & k^2 < \alpha_j^2. \end{cases} \qquad (4.37)$$

Division by $\beta_j w_j$ is a weighting that eases computations subsequently. In the next lemma, we prove that H is a compact operator from ℓ^2 into $L^2(\Omega_h)$. Therefore we introduce the sequence

$$w_j^* = \begin{cases} \exp(-\mathrm{i}\beta_j h), & k^2 > \alpha_j^2, \\ 1, & k^2 < \alpha_j^2. \end{cases}$$

Lemma 4.2.7. $H : \ell^2 \to L^2(\Omega_h)$ *is compact and injective and its adjoint* $H^* : L^2(\Omega_h) \to \ell^2$ *satisfies* $H^*(f) = -4\pi\mathrm{i}(w_j^* \hat{u}_j)_{j \in \mathbb{Z}}$ *where* $u \in H_{\alpha,0}^1(\Omega_h)$ *solves* $\Delta u + k^2 u = f$ *in* Ω_h *subject to the Rayleigh expansion condition.*

Proof. We formally compute

$$\int_{\Omega_h} H(a_j)\overline{f}\,dx = \sum_{j\in\mathbb{Z}} \frac{a_j}{\beta_j w_j} \int_{\Omega_h} u_j^i \overline{f} dx = \left\langle (a_j), \left(\int_{\Omega_h} f\overline{\left(\frac{u_j^i}{\beta_j w_j}\right)}dx\right)\right\rangle_{\ell^2}.$$

Note that $\overline{\alpha_j} = \alpha_j$ and $\overline{\beta_j} = \beta_j$ if $k^2 > \alpha_j^2$ but $\overline{\beta_j} = -\beta_j$ else (the case $\alpha_j^2 = k^2$ is excluded by assumption in this entire section). The numbers w_j are real. Therefore

$$\overline{\left(\frac{u_j^i}{\beta_j w_j}\right)} = \begin{cases} \dfrac{1}{\beta_j w_j}\left(e^{-\mathrm{i}(\alpha_j x_1 - \beta_j x_2)} - e^{-\mathrm{i}(\alpha_j x_1 + \beta_j x_2)}\right), & k^2 > \alpha_j^2 \\[2mm] -\dfrac{1}{\beta_j w_j}\left(e^{-\mathrm{i}(\alpha_j x_1 + \beta_j x_2)} - e^{-\mathrm{i}(\alpha_j x_1 - \beta_j x_2)}\right), & k^2 < \alpha_j^2 \end{cases}$$

$$= \frac{1}{\beta_j w_j}\left(e^{-\mathrm{i}(\alpha_j x_1 - \beta_j x_2)} - e^{-\mathrm{i}(\alpha_j x_1 + \beta_j x_2)}\right).$$

Assume now that $u \in H^1_{\alpha,0}(\Omega_h)$ is a weak solution to $\Delta u + k^2 u = f$ in Ω_h, $u = 0$ on Γ_0, and $\partial_2 u = T^+(u)$ on Γ_h. The variational formulation of this problem is

$$\int_{\Omega_h} \left(\nabla u \cdot \nabla\overline{v} - k^2 uv\right) dx - \int_{\Gamma_h} \overline{v}T^+(u)\,ds = -\int_{\Omega_h} f\overline{v}\,dx \quad \text{for all } v \in H^1_{\alpha,0}(\Omega_h).$$

This problem is uniquely solvable for any wave number $k > 0$. As in Section 4.1.3 one shows that the sesquilinear form satisfies a Gårding inequality, and thus all what remains to show is uniqueness of solution. By interior elliptic regularity results [39], any solution $u \in H^1_{\alpha,0}(\Omega_h)$ for $f = 0$ belongs to $H^2(\Omega_h)$. Plugging in $\partial_2 u$ into the variational formulation with $f = 0$, and taking twice the real part yields by partial integration that

$$0 = 2\Re \int_{\Omega_h} \left(\nabla u \cdot \nabla \partial_2 \overline{u} - k^2 u \partial_2 \overline{u}\right) dx - 2\Re \int_{\Gamma_h} \partial_2 \overline{u} T^+(u)\,ds$$

$$= \int_{\Omega_h} \left(\partial_2 |\nabla u|^2 - k^2 \partial_2 |u|^2\right) dx - 2\int_{\Gamma_h} |\partial_2 u|^2\,ds$$

$$= \int_{\Gamma_h} |\nabla u|^2\,ds - \int_{\Gamma_0} |\nabla u|^2\,ds - k^2 \int_{\Gamma_h} |u|^2\,ds - 2\Re \int_{\Gamma_h} |\partial_2 \overline{u}|^2\,ds$$

$$= \int_{\Gamma_h} \left(|\partial_1 u|^2 - |\partial_2 u|^2 - k^2 |u|^2\right) ds - \int_{\Gamma_0} |\partial_2 u|^2\,ds. \quad (4.38)$$

Since $\partial_2 u = T^+(u)$ we find through the expansion $u|_{\Gamma_h} = \sum_{j\in\mathbb{Z}} \hat{u}_j \exp(\mathrm{i}(\alpha_j x_1))$ that

$$|\partial_1 u|^2 - |\partial_2 u|^2 - k^2 |u|^2 = 2\pi \sum_{j\in\mathbb{Z}} |\hat{u}_j|^2 \left(\alpha_j^2 - |\alpha_j^2 - k^2| - k^2\right) \le 0$$

and, hence, (4.38) yields that the normal derivative $\partial_2 u$ vanishes on Γ_0. By construction, u vanishes on Γ_0, too, that is, the Cauchy data of u vanishes. Then

Holmgren's theorem [29] shows that u vanishes in Ω_h. Actually, in this simple geometric setting, this can also be deduced from the Rayleigh expansion.

The Rayleigh expansion condition implies that $u|_{\Gamma_h} = \sum_{l \in \mathbb{N}} \hat{u}_l \exp(i\alpha_l x_1)$. Thus, Green's second identity implies, for $v_j := u_j^i/(\beta_j w_j)$, that

$$0 = \int_{\Omega_h} \overline{(\Delta v_j + k^2 v_j)} u \, dx = \int_{\Omega_h} \overline{v_j} \left(\Delta u + k^2 u \right) dx + \int_{\Gamma_h} \left(u \partial_2 \overline{v_j} - \overline{v_j} \partial_2 u \right) ds$$

$$= \int_{\Omega_h} \overline{v_j} f \, dx + \sum_{l \in \mathbb{Z}} \hat{u}_l \int_{\Gamma_h} \left(e^{i\alpha_l x_1} \partial_2 \overline{v_j} - i\beta_l e^{i\alpha_l x_1} \overline{v_j} \right) ds.$$

Further,

$$\overline{v_j}|_{\Gamma_h} = \frac{1}{\beta_j w_j} (e^{i\beta_j h} - e^{-i\beta_j h}) e^{-i\alpha_j x_1}, \quad \partial_2 \overline{v_j}|_{\Gamma_h} = \frac{i}{w_j} (e^{i\beta_j h} + e^{-i\beta_j h}) e^{-i\alpha_j x_1},$$

and altogether we find

$$0 = \int_{\Omega_h} \overline{v_j} f \, dx + 2\frac{i\hat{u}_j}{w_j} \int_{\Gamma_h} e^{-i\beta_j h} ds = \int_{\Omega_h} \overline{v_j} f \, dx + \begin{cases} 4\pi i e^{-i\beta_j h} \hat{u}_j, & k^2 > \alpha_j^2, \\ 4\pi i \hat{u}_j, & k^2 < \alpha_j^2. \end{cases}$$

We have hence shown that $H^*(f) = -4\pi i (w_j^* \hat{u}_j)_{j \in \mathbb{Z}}$. Since the operations $f \mapsto u|_{\Gamma_h}$ and $u|_{\Gamma_h} \mapsto (\hat{u}_j)$ are bounded from $L^2(\Omega_h)$ into $H_\alpha^{1/2}(\Gamma_h)$ and from $H_\alpha^{1/2}(\Gamma_h)$ into ℓ^2, respectively, and since $(w_j^*)_{j \in \mathbb{Z}}$ is a bounded sequence, H^* is a bounded operator. Moreover, elliptic regularity results [39] imply that u is H^2-regular in a neighbourhood of Γ_h, thus, $f \mapsto u|_{\Gamma_h}$ is a compact operation from $L^2(\Omega_h)$ into $H_\alpha^{1/2}(\Gamma_h)$ and H^* is a compact operator. Therefore H is compact as well.

To show that H is injective it is sufficient to show that H^* has dense range, which follows from the fact that all sequences $(\delta_{jl})_{l \in \mathbb{Z}}$ belong to the range of H^* (by definition, the Kronecker symbol δ_{jl} equals one for $j = l$ and zero otherwise). To see this, we note that $\exp(i(\alpha_j x_1 + \beta_j (x_2 - h)))$ has Rayleigh sequence $(\delta_{jl})_{l \in \mathbb{Z}}$. Choose a cut-off function $\chi \in C^\infty(\mathbb{R})$ such that $\chi(t) = 0$ for $t < 0$ and $\chi(t) = 1$ for $t > h/2$. Then $(x_1, x_2) \mapsto \chi(x_2) \exp(i(\alpha_j x_1 + \beta_j (x_2 - h)))$ has Rayleigh sequence $(\delta_{jl})_{l \in \mathbb{Z}}$, too, satisfies a homogeneous Dirichlet boundary condition on Γ_0, and a Helmholtz equation with some right-hand side $f \in L^2(\Omega_h)$ that we could compute explicitly. According to the above computations, $H^*(f) = -4\pi i (w_l^* \delta_{jl})_{l \in \mathbb{Z}}$. By a simple scaling this implies that $(\delta_{jl})_{l \in \mathbb{Z}} \in \mathrm{Rg}(H^*)$ for any $j \in \mathbb{Z}$. $\square$

Next we define and factorise the data operator of the inverse transmission problem. This data operator models measurements in a finite distance from the periodic inhomogeneous medium of scattered fields caused by the incident waves (4.34). Due to the near field measurement methodology this operator is usually referred to as the near field operator, denoted by N. We define $N : \ell^2 \to \ell^2$ to map a sequence (a_j) to the Rayleigh sequence of the scattered field caused by the incident field $H(a_j)$,

$$N := GH. \tag{4.39}$$

The *inverse periodic transmission problem* is now to reconstruct the support of the contrast $q = n^2 - 1$ when given the near field operator N. We solve this problem using the Factorisation method and start again with a suitable factorisation.

Theorem 4.2.8. *The near field operator satisfies*

$$WN = \frac{ik^2}{4\pi} H^* TH,$$

where $W : \ell^2 \to \ell^2$ is defined by $W(a_j) = (w_j^* a_j)_{j \in \mathbb{Z}}$. $T : L^2(\Omega_h) \to L^2(\Omega_h)$ is defined by $Tf = q(f + v|_D)$, where $v \in H^1_{\alpha,0}(\Omega_h)$ is the radiating solution to (4.36).

Proof. Let us first note that W is an invertible operator on ℓ^2, which is clear since $w_j^* = 1$ for $k^2 < \alpha_j^2$ and $|w_j^*| = \left| \exp\left(-i\sqrt{k^2 - \alpha_j^2} \right) \right| = 1$ else. For simplicity, we denote by Q the operator that maps $f \in L^2(\Omega_h)$ to the Rayleigh sequence $(\hat{u}_j)_{j \in \mathbb{Z}}$ of $u \in H^1_{\alpha,0}(\Omega_h)$, solution to $\Delta u + k^2 u = f$. This operator already appeared in Lemma 4.2.7, where we showed that $H^* = -4\pi i W Q$, that is, $Q = i/(4\pi) W^{-1} H^*$. By definition of the solution operator G we have $Gf = (\hat{u}_j)_{j \in \mathbb{Z}}$ where $u \in H^1_{\alpha,0}(\Omega_h)$ is a radiating weak solution to $\Delta u + k^2(1+q)u = -k^2 qf$. This means that $\Delta u + k^2 u = -k^2 q(f + u)$, thus, $Gf = -k^2 Q(q(f + u))$. We substitute the above representation of Q to find that

$$Gf = -\frac{ik^2}{4\pi} W^{-1} H^* (q(f + u)),$$

which yields the claimed representation of $N = GH$.   □

With this factorisation, we can again use the abstract result from Theorem 4.2.5 on range equalities for a characterisation of the support of the periodic contrast q. There are merely two ingredients lacking: We need to show that T is an isomorphism with semi-definite imaginary part and we also need to characterise the support of q using the range of the solution operator G by the help of special test functions.

Lemma 4.2.9. *Suppose that the contrast $q \in L^\infty(\Omega_h)$ satisfies $\Re(q) \geq c > 0$ and $\Im(q) \geq 0$ and that the direct scattering problem (4.36) is uniquely solvable for any $f \in L^2(\Omega_h)$. Then the following statements hold.*

(a) *The middle operator $T : L^2(\Omega_h) \to L^2(\Omega_h)$ is injective and has a natural splitting $T = T_1 + T_0$ into a sum of a coercive operator T_0, $T_0 f := k^2 qf$, and a compact operator T_1, $T_1 f = k^2 q\, v|_D$, where v solves (4.36). Moreover, T is an isomorphism and $\Im(\int_{\Omega_h} \overline{f} Tf\, dx) \geq 0$ for $f \in L^2(\Omega_h)$.*

(b) *Suppose that the complement of $\operatorname{supp}(q)$ in Ω_h is connected and let $z \in \Omega_h$. The sequence $(r_j(z))_{j \in \mathbb{Z}}$ with*

$$r_j(z) = \frac{i}{4\pi \beta_j} e^{-i(\alpha_j z_1 + \beta_j(z_2 - h))} \tag{4.40}$$

belongs to $\operatorname{Rg}(G)$ if and only if z belongs to the interior of the support of q.

The proof of part (b) necessitates to introduce the α-quasiperiodic Green's function, which we completely avoided up to this point and also in the subsequent remainder of the section. This part of the proof is actually the only point where we need Green's functions in this section.

Proof. (a) The operator T_1 is compact since $f \mapsto v$ is a compact mapping. This follows from elliptic regularity results [39], since the solution to (4.36) belongs to $H^2(\Omega_h)$. Assume that $Tf = q(f + v) = 0$. Then we note as in the proof of Theorem 4.2.4 that v is a radiating solution to the homogeneous problem $\Delta v + k^2 v = 0$ subject to the Dirichlet condition $u = 0$ on Γ_0. However, we showed in the proof of Lemma 4.2.7 that this implies that v vanishes. Hence, f vanishes, too, which shows that T is injective. Finally, concerning the sign of the imaginary part of T the analogous computation to (4.31) shows that

$$\Im\left(\int_{\Omega_h} \overline{f}\, Tf \, dx\right) = k^2 \int_{\Omega_h} \Im(q)|w|^2 \, dx + \sum_{j:k^2>\alpha_j^2} \sqrt{k^2 - \alpha_j^2}\,|\hat{u}_j|^2 \geq 0.$$

(b) The sequence $(r_j(z))_{j\in\mathbb{Z}}$ is the Rayleigh sequence of the α-quasiperiodic Green's function

$$G(x, z) = \frac{\mathrm{i}}{4\pi} \sum_{j\in\mathbb{Z}} \frac{1}{\beta_j} e^{\mathrm{i}(\alpha_j(x_1-z_1)+\beta_j|x_2-z_2|)}, \qquad x \neq z,\ x, z \in \Omega_h.$$

Assume that $(r_j(z))_{j\in\mathbb{Z}} \in \mathrm{Rg}(G)$ for z not in the interior of $\mathrm{supp}(q)$. Then there are $u \in H^1_{\alpha,0}(\Omega_h)$ and $f \in L^2(\Omega_h)$ such that $\Delta u + k^2 n^2 u = -k^2 qf$ in the variational sense and $\hat{u}_j = r_j(z)$ for $j \in \mathbb{Z}$. Since the Rayleigh sequences of $G(\cdot, z)$ and u are equal, both functions are equal on a strip $(-\pi, \pi) \times (h - \delta, h)$ where $\delta > 0$ is chosen such that $\sup_{(x_1,x_2)^\top \in \mathrm{supp}(q)}(x_2) < h - \delta$. This equality follows from the representation (4.13) which is the same for both functions $G(\cdot, z)$ and u. Therefore, by analytic continuation, $G(\cdot, z)$ equals u in the complement of $\mathrm{supp}(q) \cup \{z\}$. However, u belongs to $H^1_{\alpha,0}(\Omega_h)$ but from [6] we know that $G(\cdot, z)$ has a logarithmic singularity at z. Therefore $G(\cdot, z) \notin H^1_{\alpha,0}(\Omega_h)$ (the gradient is not square integrable). Since $G(\cdot, z) = u$ in the complement of $\mathrm{supp}(q) \cup \{z\}$ we arrive at a contradiction. Hence, $(r_j(z))_{j\in\mathbb{Z}}$ cannot belong to $\mathrm{Rg}(G)$ for $z \notin \mathrm{supp}(q)$.

For $z \in \mathrm{supp}(q)$ we need to construct $f \in L^2(\Omega_h)$ with $G(f) = (r_j(z))_{j\in\mathbb{Z}}$. For this construction one uses the same technique as in Lemma 4.2.2, and therefore we omit the details here. $\qquad\square$

Theorem 4.2.10. *Suppose that the contrast $q \in L^\infty(\Omega_h)$ satisfies $\Re(q) \geq c > 0$ and $\Im(q) \geq 0$ and that the direct problem (4.36) is uniquely solvable. Let $\gamma = -4\pi\mathrm{i}/k^2$ and denote by $(\lambda_j, \psi_j)_{j\in\mathbb{N}}$ an orthonormal eigensystem of $(\gamma WN)_\sharp = |\Re(\gamma WN)| + \Im(\gamma WN)$ and by $r(z) = (r_l(z))_{l\in\mathbb{Z}}$ the test sequence from (4.40). A point $z \in \Omega_h$ belongs to the support of q if and only if*

$$\sum_{j=1}^\infty \frac{|\langle r(z), \psi_j\rangle|^2}{\lambda_j} < \infty. \tag{4.41}$$

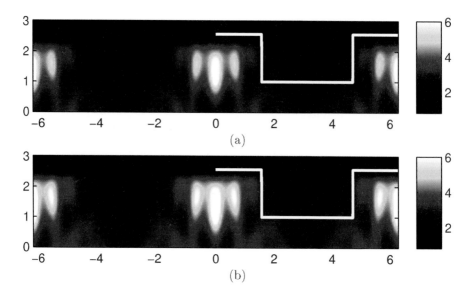

Figure 4.8: Inverse scattering for a periodic penetrable medium (see Fig. 4.2), modeling a dielectric material mounted upon a perfectly conducting plate. The plate is modeled through a Dirichlet boundary condition at Γ_0. The wave number equals 2π, the contrast inside the obstacle equals 0.5, and we use 33 incident waves. The boundary of the obstacle is indicated by a white line (over one period). The four plots show two periods of the reconstruction using a regularised truncated indicator function of the factorisation method, see (4.41). (a) No artificial noise. (b) Relative noise level of one percent.

Proof. We apply Theorem 4.2.5 to the factorisation $\gamma WN = H^*TH$ of the near field operator. The assumptions on H^* and T have been checked in Lemma 4.2.9(a) from which we obtain the identity $\mathrm{Rg}((\gamma WN)_\sharp) = \mathrm{Rg}(G) = \mathrm{Rg}(H^*T) = \mathrm{Rg}(H^*)$ since T is an isomorphism. Combination of this range identity with the characterisation given in Lemma 4.2.9(b) yields the criterion (4.41). □

Again, we finish with a numerical example (more examples can be found in [38, 5]. The boundary of the medium is given by the 2π periodic extensions to the real line of the function $f(x_1) = 1 + \pi/2 \; \mathbf{1}_{[-\pi/2,\pi/2)}$, $x_1 \in [-\pi, \pi)$, respectively. The contrast of the medium in between this boundary and the line Γ_0 equals 0.5. We use 33 incident waves and show the reconstructions using the inverse of the square root of the series in (4.41). The series index j runs from 10 to 29, which provided better results than starting the series at $j = 1$. The plot in Figure 4.2(b) is computed from data where we added 1 percent of uniformly distributed artificial noise to the synthetically computed data.

Bibliography

[1] T. Abboud. Electromagnetic waves in periodic media. In *Second International Conference on Mathematical and Numerical Aspects of Wave Propagation*, pages 1–9, Newark, DE, 1993. SIAM, Philadelphia.

[2] T. Abboud. Formulation variationelle des équations de Maxwell dans un réseau bipériodique de $\mathbf{R}^3$. *C.R. Acad. Sci. Paris, Série I*, 317:245–248, 1993.

[3] T. Abboud and J. C. Nédélec. Electromagnetic waves in an inhomogeneous medium. *J. Math. Anal. Appl.*, 164:40–58, 1992.

[4] T. Arens. Why linear sampling works. *Inverse Problems*, 20:163–173, 2004.

[5] T. Arens and N. I. Grinberg. A complete factorization method for scattering by periodic structures. *Computing*, 75:111–132, 2005.

[6] T. Arens and A. Kirsch. The factorization method in inverse scattering from periodic structures. *Inverse Problems*, 19:1195–1211, 2003.

[7] T. Arens and A. Lechleiter. The linear sampling method revisited. *J. Int. Eq. Appl.*, 21:179–202, 2009.

[8] T. Arens, A. Lechleiter, and D. R. Luke. The MUSIC algorithm as an instance of the factorization method. *SIAM J. Appl. Math.*, 70:1283-1304, 2009.

[9] I. Babuška and A. K. Aziz. Survey lectures on the mathematical foundations of the finite element method. In A. K. Aziz, editor, *The Mathematical Foundations of the Finite Element Method with Applications to Partial Differential Equations*, pages 5–359. Academic Press, New York, 1972.

[10] G. Bao. Numerical analysis of diffraction by periodic structures: TM polarization. *Numer. Math.*, 75:1–16, 1996.

[11] G. Bao. Variational approximation of Maxwell's equations in biperiodic structures. *SIAM J. Appl. Math.*, 57:364–381, 1997.

[12] A.-S. Bonnet-Bendhia and F. Starling. Guided waves by electromagnetic gratings and non-uniqueness examples for the diffraction problem. *Mathematical Methods in the Applied Sciences*, 17:305–338, 1994.

[13] G. Bruckner and J. Elschner. The numerical solution of an inverse periodic transmission problem. *Math. Meth. Appl. Sci.*, 28:757–778, 2005.

[14] F. Cakoni and D. Colton. *Qualitative Methods in Inverse Scattering Theory. An Introduction*. Springer, Berlin, 2006.

[15] P. Scott Carney and John C. Schotland. Near-field tomography. In *Inside out: inverse problems and applications*, volume 47 of *Math. Sci. Res. Inst. Publ.*, pages 133–168. Cambridge Univ. Press, Cambridge, 2003.

[16] S. N. Chandler-Wilde and P. Monk. Existence, uniqueness, and variational methods for scattering by unbounded rough surfaces. *SIAM. J. Math. Anal.*, 37:598–618, 2005.

[17] D. Colton, J. Coyle, and P. Monk. Recent developments in inverse acoustic scattering theory. *SIAM Review*, 42:396–414, 2000.

[18] D. Colton and A. Kirsch. A simple method for solving inverse scattering problems in the resonance region. *Inverse Problems*, 12:383–393, 1996.

[19] D. Colton and R. Kress. *Inverse acoustic and electromagnetic scattering theory*. Springer, 1992.

[20] D. Dobson. A variational method for electromagnetic diffraction in biperiodic structures. *Modél. Math. Anal. Numér.*, 28:419–439, 1994.

[21] D. Dobson and A. Friedman. The time-harmonic Maxwell's equations in a doubly periodic structure. *J. Math. Anal. Appl.*, 166:507–528, 1992.

[22] J. Elschner, G. Hsiao, and A. Rathsfeld. Grating profile reconstruction based on finite elements and optimization techniques. *SIAM J. Appl. Math.*, 64:525–545, 2003.

[23] J. Elschner and G. Schmidt. Diffraction of periodic structures and optimal design problems of binary gratings. Part I: Direct problems and gradient formulas. *Math. Meth. Appl. Sci.*, 21:1297–1342, 1998.

[24] H. W. Engl, M. Hanke, and A. Neubauer. *Regularization of inverse problems*. Kluwer Acad. Publ., Dordrecht, Netherlands, 1996.

[25] I. C. Gohberg and M. G. Krein. *Introduction to the theory of linear nonselfadjoint operators*, volume 18 of *Transl. Math. Monographs*. American Mathematical Society, 1969.

[26] H. Groß and A. Rathsfeld. Mathematical aspects of scatterometry – an optical metrology technique. In *Intelligent solutions for complex problems – Annual Research Report 2007*. Weierstrass Institute for Applied Analysis and Stochastics, 2007.

[27] J. Hadamard. *Lectures on the Cauchy Problem in Linear Partial Differential Equations*. Yale University Press, New Haven, 1923.

[28] M. Hanke and M. Brühl. Recent progress in electrical impedance tomography. *Inverse Problems*, 19:S65–S90, 2003.

[29] Victor Isakov. *Inverse Problems for Partial Differential Equations*. Springer, 1998.

[30] A. Kirsch. Diffraction by periodic structures. In L. Pävarinta and E. Somersalo, editors, *Proc. Lapland Conf. on Inverse Problems*, pages 87–102. Springer, 1993.

[31] A. Kirsch. *An introduction to the mathematical theory of inverse problems.* Springer, 1996.

[32] A. Kirsch. Characterization of the shape of a scattering obstacle using the spectral data of the far field operator. *Inverse Problems*, 14:1489–1512, 1998.

[33] A. Kirsch and N. I. Grinberg. *The Factorization Method for Inverse Problems.* Oxford Lecture Series in Mathematics and its Applications 36. Oxford University Press, 2008.

[34] R. Kress. *Linear Integral Equations.* Springer, 2nd edition, 1999.

[35] A. Lechleiter. A regularization technique for the factorization method. *Inverse Problems*, 22:1605–1625, 2006.

[36] A. Lechleiter. *Factorization Methods for Photonics and Rough Surface Scattering.* PhD thesis, Universität Karlsruhe (TH), Karlsruhe, Germany, 2008.

[37] A. Lechleiter. The Factorization method is independent of transmission eigenvalues. *Inverse Problems and Imaging*, 3:123–138, 2009.

[38] A. Lechleiter. Imaging of periodic dielectrics. *BIT Numer. Math.*, 50:59–83, 2010.

[39] W. McLean. *Strongly Elliptic Systems and Boundary Integral Operators.* Cambridge University Press, Cambridge, UK, 2000.

[40] R. F. Millar. The Rayleigh hypothesis and a related least-squares solution to scattering problems for periodic surfaces and other scatterers. *Radio Science*, 8:785–796, 1973.

[41] P. Monk. *Finite Element Methods for Maxwell's Equations.* Oxford Science Publications, Oxford, 2003.

[42] J.-C. Nédélec. *Acoustic and Electromagnetic Equations.* Springer, New York etc, 2001.

[43] R. Petit, editor. *Electromagnetic theory of gratings.* Springer, 1980.

[44] Lord Rayleigh. On the dynamical theory of gratings. *Proc. R. Soc. Lon. A*, 79:399–416, 1907.

[45] A. Rieder. *Keine Probleme mit Inversen Problemen.* Vieweg, 1. edition, 2003.

[46] J. Saranen and G. Vainikko. *Periodic integral and pseudodifferential equations with numerical approximation.* Springer, 2002.

[47] G. Schmidt. On the diffraction by biperiodic anisotropic structures. *Appl. Anal.*, 82:75–92, 2003.

[48] F. Schulz. On the unique continuation property of elliptic divergence form equations in the plane. *Math. Z*, 228:201–206, 1998.

[49] G. Vainikko. Fast solvers of the Lippmann-Schwinger equation. In D.E. Newark, editor, *Direct and Inverse Problems of Mathematical Physics*, Int. Soc. Anal. Appl. comput. 5, page 423, Dordrecht, 2000. Kluwer Academic Publishers.

Chapter 5

The role of the nonlinear Schrödinger equation in nonlinear optics

by Guido Schneider

We explain the role of the Nonlinear Schrödinger (NLS) equation as an amplitude equation in nonlinear optics. The NLS equation is a universal amplitude equation which can be derived via multiple scaling analysis in order to describe slow modulations in time and space of the envelope of a spatially and temporarily oscillating wave packet. It turned out to be a very successful model in nonlinear optics. Here we explain its justification by approximation theorems and its role as amplitude equation in some problems of nonlinear optics.

5.1 Introduction

The Nonlinear Schrödinger (NLS) equation

$$\partial_T A = \mathrm{i}\nu_1 \partial_X^2 A + \mathrm{i}\nu_2 A \left|A\right|^2 \tag{5.1}$$

with $T \in \mathbb{R}$, $X \in \mathbb{R}$, $\nu_1, \nu_2 \in \mathbb{R}$, and $A(X,T) \in \mathbb{C}$ is a universal amplitude equation which can be derived via multiple scaling analysis in order to describe slow modulations in time and space of the envelope of a spatially and temporarily oscillating wave packet. For instance, for the nonlinear wave equation

$$\partial_t^2 u = \partial_x^2 u - u - u^3 \qquad (x \in \mathbb{R},\ t \in \mathbb{R},\ u(x,t) \in \mathbb{R}), \tag{5.2}$$

the ansatz for its derivation is given by

$$\varepsilon \psi_{\mathrm{NLS}} = \varepsilon A\left(\varepsilon(x - c_g t), \varepsilon^2 t\right) \mathrm{e}^{\mathrm{i}(k_0 x + \omega_0 t)} + \mathrm{c.c.} \tag{5.3}$$

where $0 < \varepsilon \ll 1$ is a small perturbation parameter, where c_g is the group velocity, and where the basic temporal and spatial wave number ω_0 and k_0 are related by the linear dispersion relation $\omega_0^2 = k_0^2 + 1$. We obtain that the envelope A of the underlying carrier wave $e^{i(k_0 x + \omega_0 t)}$ has to satisfy in lowest order the NLS equation

$$2i\omega_0 \partial_T A = (1 - c_g^2)\partial_X^2 A - 3A\,|A|^2 . \tag{5.4}$$

The dynamics of the NLS equation is discussed in the subsequent section. The NLS equation possesses pulse solutions leading to modulating pulse solutions in the original system. Modulated wave packets occur in various circumstances, cf. [3], especially light pulses in nonlinear optics, or in the theory of water waves, where the NLS equation as an amplitude equation has been derived first, cf. [34]. See Figure 5.1.

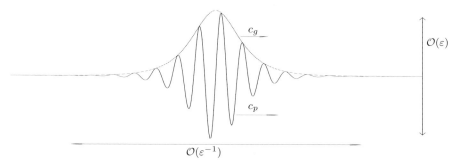

Figure 5.1: A modulating pulse described by the NLS equation. The envelope advancing with group velocity c_g in the laboratory frame modulates the underlying carrier wave advancing with group velocity c_p. The envelope evolves approximately as a solution of the NLS equation.

In this lecture notes we explain mathematical results justifying this formal derivation, i. e., we explain results that show that the NLS equation makes right predictions about the behaviour of the solutions in the original system. We explain the major steps to prove such results, i. e., we explain how to prove estimates between true solutions of the original system and the formal approximation.

In case of no quadratic terms in the original system the proof of error estimates turns out to be rather easy. The estimates follow by a simple application of Gronwall's inequality. This is the situation as it occurs in nonlinear optics due to the reflection symmetry perpendicular to the fiber. A complete proof of the estimates in this situation is given in Section 5.3. In case of quadratic terms the proof of the approximation property is presented in Section 5.4. In this case there are serious difficulties due the long $\mathcal{O}(1/\varepsilon^2)$ time scale w.r.t. t, which corresponds to an $\mathcal{O}(1)$ time scale w.r.t. T. However, by averaging or normal form techniques the quadratic terms can be eliminated and this case can be brought back to the situation discussed in Section 5.3. In Section 5.5 the content of the two previous sections is summarised and an outlook is given to related approximation questions for the NLS equation. In Section 5.6 we explain why the NLS equation is called a universal amplitude equation.

The succeeding sections are devoted to possible applications and are then written in a more informal style. We explain why the rate of information transport can be increased by multiplexing, i. e., by taking pulses with different carrier waves. We explain why photonic crystals in principle can be used as optical storage and why photonic crystals can be used as insulators perpendicular to the direction of pulse propagation.

We refer to [1] for an introduction into fiber optics from a physical point of view. There are various models claiming to be the right models for the description of nonlinear optics. Since we are interested in the mathematical ideas, we restrict ourselves to nonlinear wave equations as starting point, although these models in general are only phenomenological models of nonlinear optics, but showing dispersion and nonlinear response like optical fibers and photonic crystals.

Notation: The space of uniformly continuous and uniformly bounded functions $u : \mathbb{R} \to \mathbb{R}$ is denoted with $C_b^0(\mathbb{R}, \mathbb{R})$. It is equipped with the norm

$$\|u\|_{C_b^0} = \sup_{x \in \mathbb{R}} |u(x)|.$$

The space of m times differentiable functions $u : \mathbb{R} \to \mathbb{R}$ with uniformly continuous and uniformly bounded derivatives $\partial_x^j u$ for $j \in \{0, \ldots, m\}$ is denoted with $C_b^m(\mathbb{R}, \mathbb{R})$. It is equipped with the norm

$$\|u\|_{C_b^m} = \max_{j \in \{0, \ldots, m\}} \|\partial_x^j u\|_{C_b^0}.$$

The space of Lebesgue integrable functions $u : \mathbb{R} \to \mathbb{R}$ is denoted with $L^1(\mathbb{R}, \mathbb{R})$. It is equipped with the norm

$$\|u\|_{L^1} = \int_{\mathbb{R}} |u(x)| dx.$$

For $p \geq 1$ a function u is in $L^p(\mathbb{R}, \mathbb{R})$ if $|u|^p \in L^1$. The space L^p is equipped with the norm

$$\|u\|_{L^p} = \left(\int_{\mathbb{R}} |u(x)|^p dx \right)^{1/p}.$$

The space of m times weakly differentiable functions $u : \mathbb{R} \to \mathbb{R}$ with derivatives in L^2 for $j \in \{0, \ldots, m\}$ is denoted with $H^m(\mathbb{R}, \mathbb{R})$. This so called Sobolev space is equipped with the norm

$$\|u\|_{H^m} = \max_{j \in \{0, \ldots, m\}} \|\partial_x^j u\|_{L^2}.$$

Finally we introduce weighted Sobolev spaces $H^m(n)$ by $u \in H^m(n)$ if $u\rho^n \in H^m$ where $\rho(x) = (1 + x^2)^{1/2}$. The space $H^m(n)$ is equipped with the norm

$$\|u\|_{H^m(n)} = \|u\rho^n\|_{H^m}.$$

Acknowledgement: The author would like to thank Martina Chirilus–Bruckner, Christopher Chong, Wolf–Patrick Düll, Vincent Lescarret, and Hannes Uecker. Parts of these notes are based on joint papers or joint lecture notes.

5.2   The NLS equation

Beside the fact that the NLS equation is a universal amplitude equation which occurs in various circumstances the NLS equation is also a paradigm for a nonlinear dispersive equation which led to interesting mathematics which has been developed in relation with local and global existence questions of solutions for this equation [5, 33].

The normal form: By rescaling A, T, and X the NLS equation (5.1) can be brought into its normal form

$$\partial_T A = -\mathrm{i}\partial_X^2 A + \mathrm{i}\alpha A |A|^2$$

with $\alpha = \pm 1$. The case $\alpha = 1$ is called defocusing and the case $\alpha = -1$ is called focusing.

Local existence and uniqueness: We begin with a local existence and uniqueness result in Sobolev spaces.

Theorem 5.2.1. *Let $m \geq 1$ and $A_0 \in H^m(\mathbb{R}, \mathbb{C})$. Then there exist a time $T_0 = T_0(\|A_0\|_{H^m}) > 0$ and a unique solution $A \in C([0, T_0], H^m)$ of the NLS equation with $A|_{t=0} = A_0$.*

Proof. Since m is sufficiently large, the standard proof for semilinear evolution equations applies [20]. The operator $-\mathrm{i}\partial_X^2$ generates a strongly continuous (contraction) semigroup $(e^{-\mathrm{i}T\partial_X^2})_{T \geq 0}$ in every H^m-space as can be seen by considering the problem in Fourier space. Moreover, the nonlinearity $A \mapsto \mathrm{i}\alpha|A|^2 A$ is locally Lipschitz continuous in H^m for $m \geq 1$. The local existence and uniqueness thus follows by considering the variation of constant formula

$$A(\cdot, T) = e^{-\mathrm{i}T\partial_X^2} A(\cdot, 0) + \int_0^T e^{-\mathrm{i}(T-\tau)\partial_X^2} \mathrm{i}\alpha A(\cdot, \tau)|A(\cdot, \tau)|^2 d\tau$$

and using the contraction mapping principle. For $T_0 > 0$ sufficiently small the right hand side is a contraction in the space

$$\{A \in C([0, T_0], H^m) \mid \sup_{T \in [0, T_0]} \|A(\cdot, T)\|_{H^m} \leq 2\|A_0\|_{H^m}\}$$

equipped with norm $\|A\| = \sup_{T \in [0, T_0]} \|A(\cdot, T)\|_{H^m}$. $\square$

Dispersion: The linear problem shows dispersion. Since the linear dispersion relation for solutions $e^{ikx+i\omega t}$ is given by $\omega = k^2$ the group velocity $c_g = 2k$ varies with k. As a consequence harmonic waves move with different velocities and therefore localised wave packets dissolve. The effect of dispersion can be seen with the explicit solution formula [28, p. 4] for the solutions of the linear Schrödinger equation

$$\partial_T A = -i\partial_X^2 A$$

which is given by

$$A(X,T) = \frac{1}{\sqrt{4\pi iT}} \int_{-\infty}^{\infty} e^{-\frac{(X-Y)^2}{4iT}} A(Y,0)dY.$$

We immediately obtain the estimate

$$\sup_{X\in\mathbb{R}} |A(X,T)| \leq \frac{1}{\sqrt{4\pi T}} \int_{-\infty}^{\infty} |A(X,0)|dX,$$

i. e., solutions to spatially localised initial conditions decay uniformly towards zero with a rate $T^{-1/2}$. Dispersion conserves energy, i. e., $\frac{d}{dT}\int |A(X,T)|^2 dX = 0$, but spreads it all over the real axis.

The pulse solutions: In the focusing case, i. e. $\alpha = -1$, the focusing of energy by the nonlinearity and the defocusing of energy by dispersion are in some equilibrium such that pulse solutions of the form

$$A(X,T) = B(X)e^{i\omega T}.$$

can exist. Then B, with $B(X) \in \mathbb{R}$, satisfies

$$i\omega B = -iB'' + i\alpha B^3$$

or equivalently

$$0 = B'' + \omega B - \alpha B^3. \tag{5.5}$$

For $\alpha = -1$ and $\omega < 0$ equation (5.5) possesses two homoclinic solutions to the origin in the (B, B')-plane. See Figure 5.2. These pulse solutions correspond to the envelopes of the modulated pulse solutions of the original system drawn in Figure 5.1.

Soliton property: Due to fact that the NLS equation is a completely integrable Hamiltonian system, see below, the pulse solutions are also called solitons. There exist inverse scattering schemes, namely the AKNS–scheme and the ZS–scheme which allow to solve the NLS equation explicitly, cf. [3, 10]. A non-trivial solution which can be constructed in this way out of two single solitons is called 2-soliton. For $T \to \pm\infty$ it consists of two single pulse solutions. There is a nonlinear interaction of the solitons but except of two phase shifts after the interaction

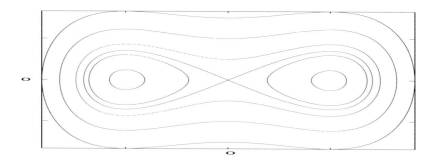

Figure 5.2: The (B, B')-phase portrait for 5.5 with the homoclinic solutions.

asymptotically the two solitons look exactly as before the interaction. Completely integrable Hamiltonian systems possess infinitely many conservation laws. As an example the L^2-norm is conserved along solutions. We have

$$(d/dT)\|A\|_{L^2}^2 = (d/dT)\int A\bar{A}\,\mathrm{d}X = \int (\partial_T A)\bar{A} + A(\partial_T \bar{A})dX$$

$$= 2\mathrm{Re}\int \bar{A}\left(-\mathrm{i}(\partial_X^2 A - \alpha|A|^2 A)\right)dX = 0.$$

In Hamiltonian systems also the Hamiltonian is conserved which is given here by

$$H(A) = \int_{\mathbb{R}} \frac{1}{2}|\partial_X A(X)|^2 + \frac{1}{4}\alpha|A(X)|^4 \,\mathrm{d}X.$$

In order to show that the NLS equation is a Hamiltonian system we first compute

$$\partial_A H[B] = \lim_{\varepsilon \to 0} \varepsilon^{-1}(H(A + \varepsilon B) - H(A))$$

$$= \lim_{\varepsilon \to 0} \varepsilon^{-1}\int_{\mathbb{R}} |\partial_X (A + \varepsilon B)|^2/2 + \alpha|A + \varepsilon B|^4/4 - |\partial_X A|^2/2 - |A|^4/4\,\mathrm{d}X$$

$$= \Re \int_{\mathbb{R}} (\partial_X^2 \bar{A} + \alpha\bar{A}|A|^2)B\,\mathrm{d}X.$$

$\partial_A H[\cdot]$ maps $B \in L^2(\mathbb{R}, \mathbb{C})$ linearly into $\mathbb{C}$ and hence $\partial_A H[\cdot]$ is an element of the dual space. By the Riesz theorem [2, Satz 4.1] a Hilbert space can be identified with its dual space via the mapping

$$\beta : \mathrm{Lin}(L^2, \mathbb{R}) \to L^2, \qquad \left(B \mapsto \langle A, B\rangle = \Re \int \bar{A}(X)B(X)\,\mathrm{d}X\right) \mapsto A.$$

Hence $\beta\partial_A H = \partial_X^2 A + \alpha A|A|^2$ and so finally the NLS equation can be written as abstract Hamiltonian system

$$\partial_T A = -\mathrm{i}\partial_X^2 A + \alpha\mathrm{i}A|A|^2 = \mathrm{i}\beta\partial_A H(A) = J\beta\partial_A H(A)$$

with the skew symmetric operator $(JA) = iA$. The skew symmetry of J in L^2 follows from

$$\langle JA, B \rangle = \Re \int \overline{iA(X)}B(X)\,dX = -\Re \int \overline{A(X)}iB(X)\,dX = -\langle A, JB \rangle.$$

In the defocusing case the Hamiltonian is positive definite, i. e., $H(A) > 0$ for $A \neq 0$, whereas in the focusing case it is indefinite. Hence, in the defocusing case we can use the two conserved quantities, the L^2-norm and the Hamiltonian H to obtain the global existence and uniqueness of solutions in H^1. Due to the local existence and uniqueness in H^1 an a priori estimate for the H^1-norm is sufficient for this result. We get the H^1-estimate for free since

$$\|A(T)\|_{H^1} \leq H(A(T)) + \|A(T)\|_{L^2} = H(A(0)) + \|A_0\|_{L^2}^2$$

for all $T \in \mathbb{R}$.

5.3 Cubic Nonlinearities

In this section we explain how to justify the NLS equation in case of cubic non-linearities in the original system. For expository reasons we restrict ourselves to

$$\partial_t^2 u = \partial_x^2 u - u - u^3 \tag{5.6}$$

with $x \in \mathbb{R}$, $t \in \mathbb{R}$ and $u(x,t) \in \mathbb{R}$ as original system. Our main purpose is to prove that solutions of (5.6) behave as predicted by the associated NLS equation. The NLS approximation

$$\varepsilon\psi_{\mathrm{NLS}} = \varepsilon A\left(\varepsilon(x - c_g t), \varepsilon^2 t\right) e^{i(k_0 x + \omega_0 t)} + \mathrm{c.c.} \tag{5.7}$$

is formally a good approximation if the terms which do not cancel after inserting $\varepsilon\psi_{\mathrm{NLS}}$ into (5.6) are small.

The residual: These terms are collected in the so-called residual

$$\mathrm{Res}(u) = -\partial_t^2 u + \partial_x^2 u - u - u^3. \tag{5.8}$$

If $\mathrm{Res}(u) = 0$ then u is an exact solution of (5.6). With the abbreviation $E = e^{i(k_0 x + \omega_0 t)}$ we find

$$
\begin{aligned}
\mathrm{Res}(\varepsilon\psi_{\mathrm{NLS}}) \;=\; & \varepsilon E\left((\omega_0^2 - k_0^2 - 1)A\right) \\
& + \varepsilon^2 E\left((2ik_0 - 2ic_g\omega_0)\partial_X A\right) \\
& + \varepsilon^3 E\left((-2i\omega_0\partial_T A + (1 - c_g^2)\partial_X^2 A - 3A|A|^2\right) \\
& + \varepsilon^3 E^3(-A^3) \\
& + \varepsilon^4 E(2c_g\partial_X\partial_T A) \\
& + \varepsilon^5 E(-\partial_T^2 A) + \mathrm{c.c.}
\end{aligned}
$$

By choosing $\omega = \omega_0$ and $k = k_0$ to satisfy the linear dispersion relation

$$\omega^2 = k^2 + 1,$$

by choosing c_g to be the linear group velocity

$$c_g = \left.\frac{d\omega}{dk}\right|_{k=k_0} = \frac{k_0}{\omega_0}$$

and by choosing A to satisfy the NLS equation

$$2i\omega_0\partial_T A = (1 - c_g^2)\partial_X^2 A - 3A\,|A|^2 \tag{5.9}$$

the first three lines in the residual cancel.

Unfortunately, we still have $\mathrm{Res}(\varepsilon\psi_{\mathrm{NLS}}) = \mathcal{O}(\varepsilon^3)$ (why we need a smaller residual will become clear in a moment).

Formal smallness of the residual: However, it turns out that by adding higher order terms to the approximation $\varepsilon\psi_{\mathrm{NLS}}$ the residual can be made arbitrarily small, i. e., for arbitrary, but fixed β there exists an approximation $\varepsilon\psi_\beta$ with $\varepsilon\psi_\beta - \varepsilon\psi_{\mathrm{NLS}} = \mathcal{O}(\varepsilon^3)$ and $\mathrm{Res}(\varepsilon\psi_\beta) = \mathcal{O}(\varepsilon^\beta)$.

Since $\varepsilon\psi_\beta - \varepsilon\psi_{\mathrm{NLS}} = \mathcal{O}(\varepsilon^3)$ the approximation $\varepsilon\psi_\beta$ makes the same predictions as $\varepsilon\psi_{\mathrm{NLS}}$ about the behaviour of the solutions u of the original system. We will show $\mathrm{Res}(\varepsilon\psi_\beta) = \mathcal{O}(\varepsilon^\beta)$ for $\beta = 4, 5$. With these two examples the general situation can be understood.

In order to obtain

$$\mathrm{Res}(\varepsilon\psi_4) = \mathcal{O}(\varepsilon^4) \tag{5.10}$$

we define

$$\varepsilon\psi_4 = \varepsilon\psi_{\mathrm{NLS}} + \left(\varepsilon^3 A_3\left(\varepsilon(x - c_g t), \varepsilon^2 t\right) E^3 + \mathrm{c.c.}\right).$$

We find

$$\mathrm{Res}(\varepsilon\psi_4) = \varepsilon^3 E^3\left(-A^3 - (9\omega_0^2 - 9k_0^2 - 1)A_3\right) + \mathcal{O}(\varepsilon^4).$$

Since $9\omega_0^2 - 9k_0^2 - 1 = 9(k_0^2 + 1) - 9k_0^2 - 1 = 8 \neq 0$ we can choose

$$A_3 = -(9\omega_0^2 - 9k_0^2 - 1)^{-1}A^3$$

in order to achieve (5.10). In order to achieve

$$\mathrm{Res}(\varepsilon\psi_5) = \mathcal{O}(\varepsilon^5) \tag{5.11}$$

we define

$$\varepsilon\psi_5 = \varepsilon\psi_4 + \left(\varepsilon^2 A_{12}(\varepsilon(x - c_g t), \varepsilon^2 t)E + \varepsilon^4 A_{32}(\varepsilon(x - c_g t), \varepsilon^2 t)E^3 + \mathrm{c.c.}\right).$$

where A_{12} and A_{32} are new functions to be chosen below. We find

$$
\begin{aligned}
\mathrm{Res}(\varepsilon\psi_5) \;=\; & \varepsilon^4 E\left(-2\mathrm{i}\omega_0\partial_T A_{12} + (1-c_g^2)\partial_X^2 A_{12}\right. \\
& \left. -\,3A^2\overline{A_{12}} - 6\,|A|^2\,A_{12} - 2c_g\partial_X\partial_T A\right) \\
& +\,\varepsilon^4 E^3\big((9\omega_0^2 - 9k_0^2 - 1)A_{32} - 3A^2 A_{12}\big) + \mathcal{O}(\varepsilon^5) + \text{c.c.}
\end{aligned}
$$

By choosing A_{12} to satisfy the linearised NLS equation

$$
-2\mathrm{i}\omega_0\partial_T A_{12} + (1-c_g^2)\partial_X^2 A_{12} - 3A^2\overline{A_{12}} - 6\,|A|^2\,A_{12} - 2c_g\partial_X\partial_T A = 0
$$

and A_{32} to satisfy

$$
(9\omega_0^2 - 9k_0^2 - 1)A_{32} - 3A^2 A_{12} = 0
$$

we achieve (5.11). In order to achieve

$$
\mathrm{Res}(\varepsilon\psi_\beta) = \mathcal{O}(\varepsilon^\beta)
$$

we choose

$$
\varepsilon\psi_\beta = \sum_{|m|=1,3,\ldots,2N+1}\;\sum_{n=1}^{\tilde{\beta}(m)} \varepsilon^{\alpha(m)+n} A_{mn}(X,T)E^m
$$

with N and $\tilde{\beta}(m)$ sufficiently big, where

$$
\begin{aligned}
\alpha(m) &= \big|\,|m|-1\,\big|, \\
X &= \varepsilon(x - c_g t), \\
T &= \varepsilon^2 t.
\end{aligned}
$$

As before A_{11} satisfies the NLS equation, the A_{1n} for $n \geq 2$ linearised NLS equations, and the A_{mn} for $m \neq \pm 1$ algebraic equations.

Estimates for the residual: These formal orders of the residual can be improved to estimates in norms. We find for instance

$$
\|\mathrm{Res}(\varepsilon\psi_{\mathrm{NLS}})\|_{C_b^0} \leq s_1 + s_2 + s_3
$$

where

$$
\begin{aligned}
s_1 &= 2\|\varepsilon^3 E^3 A^3\|_{C_b^0} \leq 2\varepsilon^3\,\|A\|_{C_b^0}^3, \\
s_2 &= 4\|\varepsilon^4 E c_g \partial_X\partial_T A\|_{C_b^0} \leq 4\varepsilon^4 c_g\|\partial_T A\|_{C_b^1}, \\
s_3 &= 2\|\varepsilon^5 E \partial_T^2 A\|_{C_b^0} \leq 2\varepsilon^5\|\partial_T^2 A\|_{C_b^0}.
\end{aligned}
$$

We can use the right hand side of the NLS equation to estimate $\|\partial_T A\|_{C_b^1}$ and $\|\partial_T^2 A\|_{C_b^0}$. For instance we have

$$
\|\partial_T A\|_{C_b^1} \leq \frac{1}{2\omega_0}\left((1-c_g^2)\|\partial_X^2 A\|_{C_b^1} + 3\|A\|_{C_b^1}^3\right) = \mathcal{O}\left(\|A\|_{C_b^3}\right).
$$

Analogously, we find
$$\|\partial_T^2 A\|_{C_b^0} = \mathcal{O}\left(\|A\|_{C_b^4}\right)$$

such that the following holds:

Proposition 5.3.1. *Let $A \in C([0, T_0], C_b^4)$ be a solution of the NLS equation. Then for all $\varepsilon_0 \in (0, 1]$ there exists a $C > 0$ such that for all $\varepsilon \in (0, \varepsilon_0)$:*

$$\sup_{0 \le t \le T_0/\varepsilon^2} \|\mathrm{Res}(\varepsilon\psi_{\mathrm{NLS}}(t))\|_{C_b^0} \le C\varepsilon^3.$$

Similarly we can prove that for every $\beta > 0$ there exists an $m \in \mathbb{N}$ sufficiently large and an approximation $\varepsilon\psi_\beta$ such that the following holds. Let $A \in C([0, T_0], C_b^m)$ be a solution of the NLS equation. Then for all $\varepsilon_0 \in (0, 1]$ there exists a $C > 0$ such that for all $\varepsilon \in (0, \varepsilon_0)$:

$$\sup_{0 \le t \le T_0/\varepsilon^2} \|\mathrm{Res}(\varepsilon\psi_\beta(t))\|_{C_b^0} \le C\varepsilon^\beta$$

and

$$\sup_{0 \le t \le T_0/\varepsilon^2} \|\varepsilon\psi_{NLS}(t) - \varepsilon\psi_\beta(t)\|_{C_b^0} \le C\varepsilon^2.$$

The equations for the error: Estimates for the residual, even in norms, are only a necessary condition that the NLS equation predicts correctly the behaviour of the original systems. By no means they are sufficient. The errors can sum up in time and there are a number of counterexamples, cf. [22, 13], showing that formally correctly derived amplitude equations make wrong predictions about the behaviour of the original system.

The error $\varepsilon^\beta R$, the difference between the correct solution u and the approximation $\varepsilon\psi = \varepsilon\psi_{\beta+2}$, is estimated with the help of Gronwall's inequality. The error
$$\varepsilon^\beta R = u - \varepsilon\psi$$

satisfies

$$\partial_t^2 R = \partial_x^2 R - R - 3\varepsilon^2\psi^2 R - 3\varepsilon^{\beta+1}\psi R^2 - \varepsilon^{2\beta}R^3 - \varepsilon^{-\beta}\mathrm{Res}(\varepsilon\psi).$$

Although there is local existence and uniqueness in C_b^m-spaces by the method of characteristics these spaces and this method are not suitable for obtaining estimates on the long time scale $\mathcal{O}(1/\varepsilon^2)$.

Estimates for the residual in Sobolev spaces: With this respect Sobolev spaces turn out to be more suitable. Hence, we **assume:**

Let $A \in C([0, T_0], H^{s_A})$ be a solution of the NLS equation with $s_A \ge 0$ sufficiently large.

As a first step we have to re-estimate the residual in Sobolev spaces. The arguments are less trivial than before according to the scaling properties of the L^2-norm. For $\varepsilon\psi_{\mathrm{NLS}}$ we find similar as before

$$\|\mathrm{Res}(\varepsilon\psi_{\mathrm{NLS}})\|_{H^s} \leq C\left(\varepsilon^3\|A(\varepsilon\cdot)\|_{H^s}^3 + \varepsilon^4\|\partial_X\partial_T A(\varepsilon\cdot)\|_{H^s} + \varepsilon^5\|\partial_T^2 A(\varepsilon\cdot)\|_{H^s}\right)$$
$$= \mathcal{O}\left(\varepsilon^3\|A(\varepsilon\cdot)\|_{H^{s+4}}\right).$$

However,

$$\|A(\varepsilon\cdot)\|_{L^2} = \left(\int |A(\varepsilon x)|^2\,\mathrm{d}x\right)^{1/2} = \varepsilon^{-1/2}\left(\int |A(X)|^2\,\mathrm{d}X\right)^{1/2} = \varepsilon^{-1/2}\|A\|_{L^2}$$

such that finally

$$\|\mathrm{Res}(\varepsilon\psi_{\mathrm{NLS}}(t))\|_{H^s} = \mathcal{O}\left(\varepsilon^{5/2}\|A\|_{H^{s+4}}\right).$$

Nevertheless, for the overall goal this loss of $\varepsilon^{-1/2}$ is no problem, since as before, the residual can be made arbitrary small by adding higher order terms to the approximation.

Lemma 5.3.2. *For all $\beta > 0$ and $s \geq 1$ there exists an s_A sufficiently large such that the following holds. Let $A \in C([0,T_0], H^{s_A})$ be a solution of the NLS equation. Then for all $\varepsilon_0 \in (0,1]$ there exists a $C > 0$ such that for all $\varepsilon \in (0,\varepsilon_0)$ there is an approximation $\varepsilon\psi_\beta$ with*

$$\sup_{0\leq t\leq T_0/\varepsilon^2} \|\mathrm{Res}(\varepsilon\psi_\beta(t))\|_{H^s} \leq C\varepsilon^{\beta-1/2}$$

and

$$\sup_{0\leq t\leq T_0/\varepsilon^2} \|\varepsilon\psi_{NLS}(t) - \varepsilon\psi_\beta(t)\|_{C_b^0} \leq C\varepsilon^2.$$

The functional analytic set-up: There are at least two possibilities to obtain the estimates for the error between the approximation $\varepsilon\psi$ and true solutions u of the full system, namely energy estimates and secondly the combination of semigroup theory with the variation of constant formula. Although energy estimates are more easy for cubic nonlinearities we use the second approach since it will be more useful in case of quadratic nonlinearities.

Before we do so, we recall properties of the Fourier transform which will be an essential tool in the following. The Fourier transform $\mathcal{F}$ of a function $u \in L^2(\mathbb{R},\mathbb{R})$ is given by

$$(\mathcal{F}u)(k) = \widehat{u}(k) = \frac{1}{2\pi}\int u(x)\mathrm{e}^{-ikx}\,\mathrm{d}x.$$

The inverse Fourier transform $\mathcal{F}^{-1}$ is given by

$$(\mathcal{F}^{-1}\widehat{u})(x) = u(x) = \int \widehat{u}(k)\mathrm{e}^{ikx}\,\mathrm{d}k.$$

Fourier transform is an isomorphism between $H^m(n)$ and $H^n(m)$, i. e., there are positive constants $C_{m,n}$ and $\widetilde{C}_{m,n}$ such that

$$\|u\|_{H^m(n)} \le C_{m,n}\|\widehat{u}\|_{H^n(m)} \le \widetilde{C}_{m,n}\|u\|_{H^m(n)}.$$

It is continuous from L^1 to C_b^0, i. e., there is $C = 1 > 0$ such that

$$\|u\|_{C_b^0} \le C\|\widehat{u}\|_{L^1}.$$

The space $H^m(n)$ is closed under pointwise multiplication for $m > 1/2$ and under convolution for $n > 1/2$, i. e.,

$$\|u \cdot v\|_{H^m(n)} \le C\|u\|_{H^m(n)}\|v\|_{H^m(n)}$$

for $m > 1/2$ and

$$\|\widehat{u} * \widehat{v}\|_{H^m(n)} \le C\|\widehat{u}\|_{H^m(n)}\|v\|_{H^m(n)}$$

for $n > 1/2$. Moreover, pointwise multiplication in x-space corresponds to convolution in Fourier space

$$\mathcal{F}(uv) = (\mathcal{F}u) * (\mathcal{F}v).$$

The Fourier transform of $A(\varepsilon x)e^{ik_0 x}$ is given by

$$\frac{1}{2\pi}\int A(\varepsilon x)e^{ik_0 x}e^{-ikx}\,\mathrm{d}x = \frac{1}{2\pi}\int A(\varepsilon x)e^{i(k_0-k)x}\,\mathrm{d}x$$

$$= \frac{1}{2\pi}\int \frac{1}{\varepsilon}A(X)e^{i(\frac{k_0-k}{\varepsilon})}\,\mathrm{d}X = \frac{1}{\varepsilon}\widehat{A}\left(\frac{k-k_0}{\varepsilon}\right). \qquad (5.12)$$

The equations for the error in Fourier space: In order to use semigroup theory we write the equations for the error as first order system in Fourier space. We choose $\varepsilon\psi = \varepsilon\psi_{\beta+5/2}$ and find

$$\partial_t^2 \widehat{R} = -(k^2+1)\widehat{R} - 3\varepsilon^2\widehat{\psi}^{*2} * \widehat{R} - 3\varepsilon^{\beta-1}\widehat{\psi} * \widehat{R}^{*2} - \varepsilon^{2\beta}\widehat{R}^{*3} + \varepsilon^{-\beta}\widehat{\mathrm{Res}(\varepsilon\psi)}.$$

This is written as

$$\partial_t \widehat{R}_1 = -\sqrt{k^2+1}\,\widehat{R}_2,$$

$$\partial_t \widehat{R}_2 = \sqrt{k^2+1}\,\widehat{R}_1 + \varepsilon^2\widehat{f},$$

where

$$\widehat{f} = \frac{1}{\sqrt{k^2+1}}\left(-3\widehat{\psi}^{*2} * \widehat{R} - 3\varepsilon^{\beta-3}\widehat{\psi} * \widehat{R}^{*2} - \varepsilon^{2\beta-2}\widehat{R}^{*3} + \varepsilon^{-\beta-2}\widehat{\mathrm{Res}(\varepsilon\psi)}\right).$$

This system is abbreviated in the following as

$$\partial_t \mathcal{R}(k,t) = \Lambda(k)\mathcal{R}(k,t) + \varepsilon^2 F(k,t).$$

We use the variation of constant formula

$$\mathcal{R}(k,t) = e^{\Lambda(k)t}\mathcal{R}(k,0) + \varepsilon^2 \int_0^t e^{\Lambda(k)(t-\tau)} F(k,\tau)\,\mathrm{d}\tau$$

in order to estimate the solutions of this system. Herein, $(e^{\Lambda(k)t})_{t\geq 0}$ is the semigroup generated by $\Lambda(k)$ which is defined for fixed k by

$$e^{\Lambda(k)t} = \sum_{n=0}^{\infty} (\Lambda(k)t)^n.$$

Lemma 5.3.3. *The semigroup is uniformly bounded in every $H^0(s)$, i. e., there exists a $C > 0$ such that we have $\sup_{t\in\mathbb{R}} \|e^{\Lambda t}\|_{H^0(s)\to H^0(s)} \leq C$.*

Proof. We have $\Lambda(k) = SD(k)S^{-1}$ where

$$S = \begin{pmatrix} 1 & i \\ 1 & -i \end{pmatrix} \qquad \text{and} \qquad D(k) = \begin{pmatrix} i\sqrt{k^2+1} & 0 \\ 0 & -i\sqrt{k^2+1} \end{pmatrix}$$

such that

$$e^{\Lambda(k)t} = Se^{D(k)t}S^{-1}.$$

Hence,

$$\|e^{\Lambda t}u\|_{H^0(s)} \leq \sup_{k\in\mathbb{R}} \|e^{\Lambda(k)t}\|_{\mathbb{C}^2\to\mathbb{C}^2} \|u\|_{H^0(s)}$$

and further

$$\sup_{k\in\mathbb{R}} \|e^{\Lambda(k)t}\|_{\mathbb{C}^2\to\mathbb{C}^2} \leq \|S\|_{\mathbb{C}^2\to\mathbb{C}^2} \sup_{k\in\mathbb{R}} \|e^{D(k)t}\|_{\mathbb{C}^2\to\mathbb{C}^2} \cdot \|S^{-1}\|_{\mathbb{C}^2\to\mathbb{C}^2}$$

$$\leq \|S\|_{\mathbb{C}^2\to\mathbb{C}^2} \|S^{-1}\|_{\mathbb{C}^2\to\mathbb{C}^2} < \infty. \qquad \square$$

Lemma 5.3.4. *For every $s \geq 1$ there is a $C > 0$ such that for all $\varepsilon \in (0,1]$ we have*

$$\|F\|_{H^0(s)} \leq C\left(\|R\|_{H^0(s)} + \varepsilon^{\beta-3}\|R\|_{H^0(s)}^2 + \varepsilon^{2\beta-2}\|R\|_{H^0(s)}^3 + 1\right).$$

Proof. The estimate follows from

$$\left\|\frac{1}{\sqrt{k^2+1}}\widehat{u}\right\|_{H^0(s)} \leq \|\widehat{u}\|_{H^0(s)},$$

$$\left\|\widehat{\psi} * \widehat{\psi} * \widehat{R}\right\|_{H^0(s)} \leq C\|\widehat{\psi}\|_{L^1(s)}^2 \|\widehat{R}\|_{H^0(s)},$$

$$\|\widehat{\psi} * \widehat{R} * \widehat{R}\|_{H^0(s)} \leq C\|\widehat{\psi}\|_{L^1(s)} \|\widehat{R}\|_{H^0(s)}^2,$$

$$\|\widehat{R}^{*3}\|_{H^0(s)} \leq C\|\widehat{R}\|_{H^0(s)}^3,$$

$$\|\varepsilon^{-\beta}\widehat{\mathrm{Res}(\varepsilon\psi)}\|_{H^0(s)} \leq C,$$

and finally

$$\|\widehat{\psi}\|_{L^1(s)} = 2\left\|\frac{1}{\varepsilon}A\left(\frac{\cdot - k_0}{\varepsilon}\right) + \text{h.o.t.}\right\|_{L^1(s)} \le C\left\|\frac{1}{\varepsilon}A\left(\frac{\cdot}{\varepsilon}\right)\right\|_{L^1(s)} + \text{h.o.t.}$$

$$\le C\|A\|_{L^1(s)} + \text{h.o.t.} \le C\|A\|_{H^0(s+1)} + \text{h.o.t.} \qquad\qquad \square$$

Remark 5.3.5. Note that $\|\widehat{\psi}\|_{H^0(s)} = \mathcal{O}(\varepsilon^{-1/2})$ such that $\widehat{\psi}$ has to be estimated in the $L^1(s)$-norm, resp. ψ in the C_b^s-norm in order to get ε^2 for the most dangerous term $\varepsilon^2\widehat{\psi} * \widehat{\psi} * \widehat{R}$. This power is necessary to obtain estimates on the natural time scale $\mathcal{O}(1/\varepsilon^2)$ w.r.t. t.

Using the previous lemmas shows the inequality

$$\|\widehat{R}(t)\|_{H^0(s)}$$

$$\le C\varepsilon^2 \int_0^t \left(\|\widehat{R}(\tau)\|_{H^0(s)} + \varepsilon^{\beta-3}\|\widehat{R}(\tau)\|_{H^0(s)}^2 + \varepsilon^{2\beta-2}\|\widehat{R}(\tau)\|_{H^0(s)}^3 + 1\right) d\tau$$

$$\le C\varepsilon^2 \int_0^t \left(\|\widehat{R}(\tau)\|_{H^0(s)} + 2\right) d\tau \le 2CT_0 + C\varepsilon^2 \int_0^t \|\widehat{R}(\tau)\|_{H^0(s)} d\tau,$$

which holds as long as

$$\varepsilon^{\beta-3}\|\widehat{R}(\tau)\|_{H^0(s)}^2 + \varepsilon^{2\beta-2}\|\widehat{R}(\tau)\|_{H^0(s)}^3 \le 1. \qquad\qquad (5.13)$$

Gronwall's inequality then yields:

$$\|\widehat{R}(t)\|_{H^0(s)} \le 2CT_0 e^{C\varepsilon^2 t} \le 2CT_0 e^{CT_0} = M$$

for all $t \in [0, T_0/\varepsilon^2]$. Choosing $\varepsilon_0 > 0$ such that $\varepsilon_0^{\beta-3}M^2 + \varepsilon_0^{2\beta-2}M^3 \le 1$ we have satisfied the condition (5.13) and so proved the following approximation result.

Theorem 5.3.6. *For all $\beta > 3$ and s there exists an s_A sufficiently large such that the following holds: Let $A \in C([0, T_0], H^{s_A})$ be a solution of the NLS equation (5.9). Then there exists an $\varepsilon_0 > 0$ and a $C > 0$ such that for all $\varepsilon \in (0, \varepsilon_0)$ there are solutions u of the original system (5.6) which can be approximated by $\varepsilon\psi$ which is $\mathcal{O}(\varepsilon^3)$ to $\varepsilon\psi_{NLS}$ defined in (5.7) such that*

$$\sup_{t\in[0,T_0/\varepsilon^2]} \|u(t) - \varepsilon\psi(t)\|_{H^s} < C\varepsilon^{\beta-1/2}.$$

and as a consequence

$$\sup_{t\in[0,T_0/\varepsilon^2]} \|u(t) - \varepsilon\psi_{NLS}(t)\|_{H^s} < C\varepsilon^{3/2}.$$

Remark 5.3.7. So far we have not made any remark about the existence and uniqueness of the solutions of the equations for the error. We have local existence and uniqueness of the solutions of the nonlinear wave equation (5.6) in the spaces where we proved the error estimates.

Fix $m \geq 1$ and let $(u_0, u_1) \in H^{m+1} \times H^m$. Then there exists a $t_0 > 0$ such that (5.6) possesses a unique solution $u \in C([-t_0, t_0], H^{m+1})$ with $u|_{t=0} = u_0$ and $\partial_t u|_{t=0} = u_1$. In order to construct solutions of (5.6) we use the formula

$$u(x, t) = \frac{1}{2}(u_0(x + t) + u_0(x - t)) + \frac{1}{2} \int_{x-t}^{x+t} u_1(\xi) \, d\xi$$
$$+ \int_0^t \int_{x-(t-s)}^{x+(t-s)} f(y, s) \, dy \, ds$$

with $f(x, t) = -u(x, t) - u(x, t)^3$ which is based on the solution formula for the inhomogeneous wave equation. For $t_0 > 0$ sufficiently small the right hand side $F(u)(x, t)$ is a contraction in the space $C([-t_0, t_0], H^{m+1})$. Thus there exists a unique fixed point $u^* = F(u^*)$ which is a classical solution of (5.6) if $m \geq 2$.

Like for ordinary differential equations the solutions exist until the norm of the solutions becomes infinite. By using the error estimates as a priori estimates we can guarantee that the solutions stay bounded for $t \in [0, T_0/\varepsilon^2]$ and so we can apply the local existence and uniqueness again and again to guarantee the existence and uniqueness of the solutions of the error equations which are obtained from (5.6) by a smooth change of variables.

5.4 Quadratic Nonlinearities

In this section we explain how to justify the NLS equation in case of quadratic nonlinearities. For expository reasons we restrict ourselves to

$$\partial_t^2 u = \partial_x^2 u - u + u^2 \tag{5.14}$$

with $x \in \mathbb{R}$, $t \in \mathbb{R}$ and $u(x, t) \in \mathbb{R}$ as original system. The ansatz for the derivation of the NLS equation is then given by

$$\varepsilon \psi = \varepsilon A_1 \left(\varepsilon(x - c_g t), \varepsilon^2 t \right) e^{i(k_0 x + \omega_0 t)} + \text{c.c.}$$
$$+ \varepsilon^2 A_2 \left(\varepsilon(x - c_g t), \varepsilon^2 t \right) e^{2i(k_0 x + \omega_0 t)} + \text{c.c.} \tag{5.15}$$
$$+ \varepsilon^2 A_0 \left(\varepsilon(x - c_g t), \varepsilon^2 t \right).$$

We find as before at εE the linear dispersion relation and at $\varepsilon^2 E$ the condition for linear group velocity c_g. At $\varepsilon^3 E$ we find

$$2i\omega_0 \partial_T A_1 = (1 - c_g^2)\partial_X^2 A_1 + 2A_1 A_0 + 2A_2 A_{-1}.$$

The algebraic relations which are found at $\varepsilon^2 E^0$ and $\varepsilon^2 E^2$

$$\varepsilon^2 E^0 : \quad 0 = -A_0 + 2A_1 A_{-1}$$
$$\varepsilon^2 E^2 : \quad 0 = -(-4\omega_0^2 + 4k_0^2 + 1)A_2 + A_1^2$$

can be solved with respect to A_0 and A_2 since $-4\omega_0^2 + 4k_0^2 + 1 \neq 0$. Inserting the solution for A_0 and A_2 into the equation for A_1 finally yields the NLS equation

$$2i\omega_0 \partial_T A_1 = (1 - c_g^2)\partial_X^2 A_1 + \gamma A_1 |A_1|^2 \tag{5.16}$$

with

$$\gamma = 4 + \frac{2}{-4\omega_0^2 + 4k_0^2 + 1}.$$

Like in case of cubic nonlinearities the residual

$$\text{Res}(u) = -\partial_t^2 u + \partial_x^2 u - u + u^2$$

can be made arbitrarily small by adding higher order terms, i. e., we have again

Lemma 5.4.1. *For all $\beta > 0$ and $s \geq 1$ there exists an s_A sufficiently large such that the following holds. For $A \in C([0, T_0], H^{s_A})$ and $\varepsilon_0 \in (0, 1]$ there exists a $C > 0$ such that for all $\varepsilon \in (0, \varepsilon_0)$ there is an approximation $\varepsilon\psi_\beta$ with*

$$\sup_{0 \leq t \leq T_0/\varepsilon^2} \|\text{Res}(\varepsilon\psi_\beta(t))\|_{H^s} \leq C\varepsilon^{\beta - 1/2}$$

and

$$\sup_{0 \leq t \leq T_0/\varepsilon^2} \|\varepsilon\psi_{\text{NLS}}(t) - \varepsilon\psi_\beta(t)\|_{C_b^0} \leq C\varepsilon^2.$$

In order to prove that the solution A of the NLS equation predicts the behaviour of the solutions u of the original system correctly we estimate as before the difference $\varepsilon^\beta R = u - \varepsilon\psi$ between the correct solution u and its approximation and as before we choose $\varepsilon\psi = \varepsilon\psi_{\beta+5/2}$. This difference satisfies

$$\partial_t^2 R = \partial_x^2 R - R + 2\varepsilon\psi R + \varepsilon^\beta R^2 + \varepsilon^{-\beta}\text{Res}(\varepsilon\psi).$$

Again this is written as a first order system in Fourier space

$$\partial_t \widehat{R}_1 = -\sqrt{k^2 + 1}\, \widehat{R}_2,$$
$$\partial_t \widehat{R}_2 = \sqrt{k^2 + 1}\, \widehat{R}_1 + \frac{1}{\sqrt{k^2 + 1}} \left(2\varepsilon\widehat{\psi} * \widehat{R} + \varepsilon^\beta \widehat{R}^{*2} + \varepsilon^{-\beta}\widehat{\text{Res}(\varepsilon\psi)} \right).$$

The simple argument of the last section no longer works because of the new $\mathcal{O}(\varepsilon)$-term $2\varepsilon\widehat{\psi} * \widehat{R}$. In principle this term can give some exponential growth of order $\mathcal{O}(\exp(\varepsilon t))$ which is definitely not $\mathcal{O}(1)$-bounded on the time scale of order $\mathcal{O}(1/\varepsilon^2)$. However, this term turned out to be oscillatory in time and can be eliminated with averaging or a normal form transformation such that it finally has an $\mathcal{O}(1)$-influence on the size of the solutions. This observation goes back to Kalyakin [15]. In order to eliminate this term we first write this system as a first order system

$$\partial_t R = \Lambda R + 2\varepsilon B(\psi, R) + \mathcal{O}(\varepsilon^2), \tag{5.17}$$

with Λ a symmetric linear and $B(\cdot, \cdot)$ a bilinear mapping. In Fourier space they are defined through

$$\hat{\Lambda} = \begin{pmatrix} -i\sqrt{k^2+1} & 0 \\ 0 & +i\sqrt{k^2+1} \end{pmatrix}, \qquad S = \tfrac{1}{\sqrt{2}}\begin{pmatrix} 1 & 1 \\ i & -i \end{pmatrix},$$

$$\hat{B}(\hat{U}, \hat{U}) = \tfrac{1}{\sqrt{k^2+1}} S^{-1}\tilde{B}(S\hat{U}, S\hat{U}), \qquad \tilde{B}(\hat{U}, \hat{V}) = \begin{pmatrix} 0 \\ \hat{U}_1 * \hat{V}_1 \end{pmatrix},$$

where $\hat{U} = (\hat{U}_1, \hat{U}_2)$. Next we make a near identity change of variables

$$R = w + \varepsilon Q(\psi, w) \tag{5.18}$$

with Q an autonomous bilinear mapping. This gives

$$
\begin{aligned}
\partial_t w + \varepsilon Q(\partial_t\psi, w) + \varepsilon Q(\psi, \partial_t w) &= \partial_t R = \Lambda R + 2\varepsilon B(\psi, R) + \mathcal{O}(\varepsilon^2) \\
&= \Lambda w + \varepsilon \Lambda Q(\psi, w) + 2\varepsilon B(\psi, w) + \mathcal{O}(\varepsilon^2)
\end{aligned}
$$

and so

$$\partial_t w = \Lambda w + \varepsilon[\Lambda Q(\psi, w) - Q(\Lambda\psi, w) - Q(\psi, \Lambda w) + 2B(\psi, w)] + \mathcal{O}(\varepsilon^2) \tag{5.19}$$

where we used $\partial_t\psi = \Lambda\psi + \mathcal{O}(\varepsilon^2)$. In order to eliminate the dangerous terms $2\varepsilon B(\psi, w)$ we have to find a Q such that

$$\Lambda Q(\psi, w) - Q(\Lambda\psi, w) - Q(\psi, \Lambda w) + 2B(\psi, w) = 0.$$

In Fourier space with

$$
\begin{aligned}
(\hat{B}(\hat{\psi}, \hat{w}))_j &= \sum_{m,n=1,2} \int \hat{b}^j_{mn}(k, k-l, l)\hat{\psi}_m(k-l)\hat{w}_n(l)\,\mathrm{d}l, \\
(\hat{Q}(\hat{w}, \hat{w}))_j &= \sum_{m,n=1,2} \int \hat{q}^j_{mn}(k, k-l, l)\hat{\psi}_m(k-l)\hat{w}_n(l)\,\mathrm{d}l
\end{aligned}
$$

for the j-th component of B and Q, $\omega_{1,2}(k) = \mp\sqrt{1+k^2}$ and $\hat{b}^j_{mn}, \hat{q}^j_{mn}$ some coefficients we obtain the well known relation

$$i(\omega_j(k) - \omega_m(k-l) - \omega_n(l))\hat{q}^j_{mn}(k, k-l, l) = \hat{b}^j_{mn}(k, k-l, l). \tag{5.20}$$

Since the approximation $\varepsilon\psi$ is of order $\mathcal{O}(\varepsilon)$ only close to the wave numbers $\pm k_0$, Equation (5.20) can be solved with respect to $\hat{q}_{mn}^h$ due to the validity of the non resonance condition

$$\inf_{j,m,n\in\{1,2\}} \inf_{k,l\in\mathbb{R},|k-l|\leq\delta} |(\omega_j(k) - \omega_m(k-l) - \omega_n(l))| \geq C > 0 \qquad (5.21)$$

for a $\delta > 0$ fixed. Since all kernels are globally Lipschitz, and since we have a real-valued problem the non-resonance condition (5.21) can be relaxed to

$$\inf_{j,n\in\{1,2\}} \inf_{k\in\mathbb{R}} |(\omega_j(k) - \omega_1(k_0) - \omega_n(k-k_0))| > 0, \qquad (5.22)$$

due to $\|\chi_{\{k\geq 0\}} (\omega(\cdot) - \omega(k_0))\, \varepsilon\psi(\cdot)\|_{L^1(s)} = \mathcal{O}(\varepsilon^2)$. Since

$$\sup_{j,m,n\in\{1,2\},\ k,l\in\mathbb{R}} |\hat{b}_{mn}^j(k, k-l, l)| \leq C < \infty$$

we obtain

$$\|(Q(\psi, w))\|_{H^s} \leq C\|\psi\|_{H^s}\|w\|_{H^s}.$$

Thus the transformation (5.18) can be solved with respect to w for $\varepsilon > 0$ sufficiently small. Therefore (5.19) transforms into

$$\partial_t w = \Lambda w + \mathcal{O}(\varepsilon^2) \qquad (5.23)$$

and so the proof of the approximation property from Section 5.3 then applies line by line to (5.23) if $\varepsilon\psi = \varepsilon\psi_{\beta+5/2}$ is chosen with $\beta > 2$.

5.5   Summary and outlook

The NLS approximation can be derived in various systems, including those that describe water waves, cf. [34], nonlinear optics, cf. [3], quantum theory, cf. [19], DNA double strands, cf. [14], and plasma physics, cf. [27]. In most of these systems, particularly in nonlinear optics, these approximations turn out to be very successful; agreement is generally better when experimental data is compared with NLS approximations opposed to existing estimates. However, sometimes this approximation fails, cf. [12]. Therefore, an important issue is the validity of the NLS approximation. Checking this directly with numerical experiments would be computationally demanding for $\varepsilon \to 0$, due to the large computational domains necessary. Ignoring these practical issues, numerical simulations would only provide evidence for the validity of the NLS approximation since only a finite dimensional phase space can be considered. Therefore, the validity of the NLS approximation can only be proved analytically and the proof of such estimates is a nontrivial task.

The goal of an approximation theorem for the NLS equation is to show that on an $\mathcal{O}(1/\varepsilon^2)$-scale w.r.t. t the approximation and the true solution of the original system which are both of order $\mathcal{O}(\varepsilon)$ differ only by an error $\mathcal{O}(\varepsilon^{1+\beta})$ for a $\beta > 0$ as $\varepsilon \to 0$. If quadratic terms are absent in the nonlinearity such an approximation result can easily be obtained with the help of Gronwall's inequality [17] as has been explained in Section 5.3. The case of non-resonant quadratic terms in the nonlinearity for general quasilinear hyperbolic systems has already been handled in [15]. The idea to handle such systems is to eliminate the quadratic terms using a normal form transform has been explained in Section 5.4. Gronwall's inequality can then be applied within the resulting system to obtain the required estimates. The quadratic terms can be eliminated if a non-resonance condition

$$r_{j_1 j_2 j_3}(k) = \min_{j_1 j_2 j_3} \inf_{k \in \mathbb{R}} |\omega_{j_1}(k) - \omega_{j_2}(k - k_0) - \omega_{j_3}(k_0)| > 0$$

is satisfied. Herein, the $\omega_j(k)$s are the curves of eigenvalues of the linearised original system. In [15] the case of quasilinear quadratic terms is excluded explicitly, i. e., only semilinear quadratic terms are allowed. The reason for this is that normal form transforms in quasilinear systems lead to a loss of regularity and thus local existence and uniqueness of solutions in general can no longer be proven.

In a number of papers we tried to weaken the non-resonance conditions necessary for the proof of an approximation theorem. In [25] an approximation theorem has been shown in case that the set of resonant wave numbers

$$\{\, k \mid r_{j_1 j_2 j_3}(k) = 0 \text{ for some } j_1, j_2, \text{ and } j_3 \,\}$$

is separated from the set of integer multiples of the basic wave number k_0 and that analytic solutions of the NLS equation are considered. The proof is based on the fact that the influence of the resonances is exponentially small due to the analyticity of the NLS solutions. In [24] the situation of a resonant wave number $k = 0$ (motivated by models for the water wave problem) has been handled in case that the nonlinearity also vanishes at $k = 0$.

In [26] the situation of a basic wave number k_0 resonant to wave numbers k_1 and k_2 has been considered. There is a three wave interaction (TWI) system

$$\begin{aligned}
\partial_T A_1 &= \mathrm{i}\gamma_1 \overline{A_2 A_3}, \\
\partial_T A_2 &= \mathrm{i}\gamma_2 \overline{A_1 A_3}, \\
\partial_T A_3 &= \mathrm{i}\gamma_3 \overline{A_1 A_2},
\end{aligned}$$

associated to the resonances, cf. [26, Sec. 3.3]. We proved an approximation theorem [26, Theorem 3.8] in case that the subspace $\{A_2 = A_3 = 0\}$ associated to the wave number k_0 is stable in the TWI system. The proof is based on a mixture of normal form transforms for the non-resonant wave numbers and energy estimates for the resonant wave numbers. There is also a counterexample [26, Sec. 4.1] showing that the NLS equation fails to approximate solutions in the original system in case of an unstable k_0-subspace in the associated TWI system and periodic boundary conditions in the original system. In [11] the ideas of [24] and [26] are brought together to handle Boussinesq equations which model the water wave problem in case of small positive surface tension.

5.6   The universality of the NLS equation

As already explained the NLS approximation can be derived in various systems. In order to explain why this is the case and why the NLS equation plays such an important role, we review the derivation of the NLS equation from (5.6) from a different point of view. This derivation will explain why the NLS equation occurs as universal amplitude equation describing the evolution of modulated wave packets. The underlying system will condense in the values of the coefficients ν_1 and ν_2 in (5.1).

The Fourier transformed system: It turns out that Fourier transform is the key for the understanding of the universality. Hence we consider (5.6) in Fourier space. The Fourier transform $\hat{u}$ satisfies

$$\partial_t^2 \hat{u}(k,t) = -\omega^2(k)\hat{u}(k,t) - \hat{u}^{*3}(k,t), \tag{5.24}$$

where $\omega(k) = \sqrt{k^2+1}\,\operatorname{sign}(k)$. By introducing $\hat{w}(k) = (\hat{u}(k), \frac{1}{\omega(k)}\partial_t\hat{u}(k))$ we rewrite (5.24) into the first order system

$$\partial_t \hat{w}(k,t) = \hat{M}(k)\hat{w}(k,t) + \hat{N}(\hat{w})(k,t), \tag{5.25}$$

where

$$\hat{M}(k) = \begin{pmatrix} 0 & \omega(k) \\ -\omega(k) & 0 \end{pmatrix}, \quad \hat{N}(\hat{w})(k,t) = \begin{pmatrix} 0 \\ \frac{-1}{\omega(k)}\hat{u}^{*3}(k,t) \end{pmatrix}.$$

This system is diagonalised for fixed wave number k. By chance for (5.25) the associated transformation $\hat{U} = \frac{1}{\sqrt{2}}\begin{pmatrix} 1 & 1 \\ i & -i \end{pmatrix}$ is independent of k and unitary, i. e. $\hat{U}^{-1} = \hat{U}^*$. The transformed variable $\hat{z} = \hat{U}^*\hat{w}$ satisfies the diagonalised system

$$\partial_t \hat{z} = \hat{\Lambda}\hat{z} + \hat{U}^*\hat{N}(\hat{U}\hat{z}), \tag{5.26}$$

with $\hat{\Lambda}(k) = \operatorname{diag}(i\omega(k), -i\omega(k))$.

Derivation of the NLS equation: According to (5.12) for (5.26) we make the ansatz

$$
\begin{aligned}
\hat{z}(k,t) \;=\; & \varepsilon\varepsilon^{-1}\hat{A}_1\left(\frac{k-k_0}{\varepsilon}, \varepsilon^2 t\right) e^{i\omega(k_0)t} e^{ic_g(k-k_0)t}\vec{e}_1 \\
& + \varepsilon\varepsilon^{-1}\hat{A}_{-1}\left(\frac{k+k_0}{\varepsilon}, \varepsilon^2 t\right) e^{-i\omega(k_0)t} e^{ic_g(k+k_0)t}\vec{e}_2,
\end{aligned}
\tag{5.27}
$$

cf. (5.12), where

$$\vec{e}_1 = \begin{pmatrix} 1 \\ 0 \end{pmatrix} \quad \text{and} \quad \vec{e}_2 = \begin{pmatrix} 0 \\ 1 \end{pmatrix}.$$

Since the Fourier modes of the wave packet are concentrated in an $\mathcal{O}(\varepsilon)$ neighbourhood of the basic wave numbers $\pm k_0$ the evolution of the wave packet will be strongly determined by the curves $\pm\omega$ at $\pm k_0$. At $\mathrm{e}^{\mathrm{i}\omega(k_0)t}\mathrm{e}^{\mathrm{i}c_g(k-k_0)t}\vec{e}_1$ we find

$$\mathrm{i}\omega(k_0)\hat{A}_1 + \mathrm{i}\varepsilon c_g K \hat{A}_1 + \varepsilon^2 \partial_T \hat{A}_1 = \mathrm{i}\omega(k_0)\hat{A}_1 + \mathrm{i}\varepsilon \partial_k \omega(k_0) K \hat{A}_1$$

$$+ \frac{\mathrm{i}}{2}\varepsilon^2 \partial_k^2 \omega(k_0) K^2 \hat{A}_1 + \varepsilon^2 \frac{3i}{4\omega(k_0)}\hat{A}_1 * \hat{A}_1 * \hat{A}_{-1} + \mathcal{O}(\varepsilon^3),$$

where $k = k_0 + \varepsilon K$ and where we used

$$\int \hat{A}\left(\frac{k-\ell-k_0}{\varepsilon}\right)\mathrm{e}^{\mathrm{i}\omega(k_0)t}\mathrm{e}^{\mathrm{i}c_g(k-\ell-k_0)t}\hat{A}\left(\frac{\ell-k_0}{\varepsilon}\right)\mathrm{e}^{\mathrm{i}\omega(k_0)t}\mathrm{e}^{\mathrm{i}c_g(\ell-k_0)t}\,\mathrm{d}\ell$$

$$= \varepsilon\int\hat{A}\left(\frac{k-2k_0}{\varepsilon}-m\right)\hat{A}(m)\mathrm{e}^{2\mathrm{i}\omega(k_0)t}\mathrm{e}^{\mathrm{i}c_g(k-2k_0)t}\,\mathrm{d}m\ .$$

At ε^0 and ε^1 we obtain the linear dispersion relation and the linear group velocity. At ε^2 we obtain an NLS equation.

The general situation: By this procedure it is clear that the NLS equation occurs as amplitude equation of dispersive systems whenever the Fourier transform of the initial condition is strongly concentrated at a wave number $k_0 \neq 0$ and when the concentration and the amplitude are of the correct order.

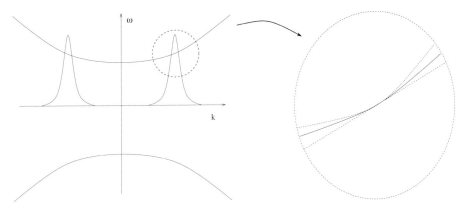

Figure 5.3: Derivation of the NLS equation is based on the concentration of the Fourier modes at a certain wave number. The left panel shows the curves of eigenvalues and the concentration of the Fourier modes. Hence for the evolution of these modes only the curves of eigenvalues close to these wave numbers plays a role. The right panel therefore shows an expansion of these curves at these wave numbers.

Therefore, the NLS equation is the universal amplitude equation describing slow modulations in time and space of a propagating wave packet in dispersive systems

$$\partial_t \hat{z}_j(k,t) = \mathrm{i}\omega_j(k)\hat{z}_j(k,t) + \text{ nonlinear terms} \tag{5.28}$$

for j in some index set. An example is the one dimensional Maxwell–Lorentz system

$$\partial_x^2 u = \partial_t^2 u + \partial_t^2 p, \qquad \partial_t^2 p + \omega_0^2 p + rp|p|^2 = du,$$

with coefficients $\omega_0, r, d \in \mathbb{R}$, describing the electric field u and some polarisation field p, where the polarisation field p is modeled as a forced nonlinear oscillator. The Fourier transformed system is written as first order system and then diagonalised leading to

$$\partial_t \hat{z}_1(k,t) \quad = \quad \mathrm{i}\omega_1(k)\hat{z}_1(k,t) + \text{ nonlinear terms },$$

$$\vdots$$

$$\partial_t \hat{z}_4(k,t) \quad = \quad \mathrm{i}\omega_4(k)\hat{z}_4(k,t) + \text{ nonlinear terms },$$

with curves of eigenvalues plotted in Figure 5.4. Hence the NLS equation can also be derived for this more realistic model of nonlinear optics. The justification is not completely trivial due to the occurrence of some Jordan block at $k = 0$, cf. [7].

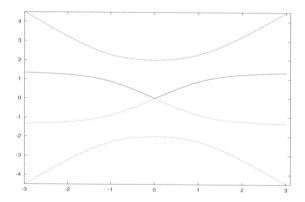

Figure 5.4: The curves of eigenvalues for the Maxwell–Lorentz system.

5.7 Applications

In this section we use the NLS equation to give some insight into a number of problems from nonlinear optics. This will be done in a rather informal style. We explain the possibility of multiplexing and the use of photonic crystals as optical storage and as wave guides.

5.7.1 Multiplexing

So far we considered pulses modulating one carrier wave with some basic wave number, say k_0. Now two or more carrier waves with different basic wave numbers, say $k_1, \ldots, k_N$ are taken. Since it turns out that pulses to different carrier waves do not interact in lowest order the use of more than one carrier wave allows to increase the rate of information through one fiber. This procedure is called multiplexing. See Figure 5.5. In the following we explain these ideas with the help of the NLS approximation.

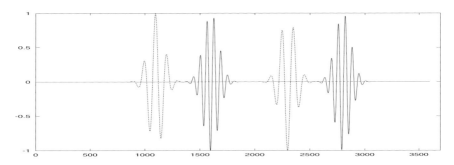

Figure 5.5: Multiplexing: The rate of information through the fiber is increased by using the same fiber for sending digital information in different bands, i. e. physically wave packets with different carrier waves are sent. In each band independently 0s and 1s are sent by sending a light pulse with this carrier wave or not. Why this works, i. e. why the pulses to different carrier waves do not interact in lowest order although they travel through each other can be explained via the NLS approximation as will become clear in this section.

For simplicity let us consider here again the nonlinear wave equation with cubic nonlinearity

$$\partial_t^2 u = \partial_x^2 u - u - u^3$$

as original system and let us consider the situation of N different carrier waves. By making the ansatz

$$\varepsilon \psi_{\text{multiNLS}} = \sum_{j=1}^{N} \varepsilon A_j \left(\varepsilon(x - c_j t), \varepsilon^2 t \right) \mathrm{e}^{\mathrm{i}(k_j x + \omega_j t)} + \text{c.c.}$$

a system of coupled NLS equations is obtained, namely

$$2 \mathrm{i} \omega_j \partial_T A_j = (1 - c_j^2) \partial_{X_j}^2 A_j - 3 A_j |A_j|^2 - \text{coupling terms}_j$$

for $j = 1, \ldots, N$, where the jth coupling term is given by

$$6 \sum_{|n| = 1, \ldots, N, n \neq j} A_j |A_n|^2,$$

where we assumed for simplicity that there are no additional indices $j_1, \ldots, j_4$ with $k_{j_1} + \ldots + k_{j_4} = 0$. If this assumption is not satisfied there will be additional coupling terms $A_{j_2} A_{j_3} A_{j_4}$ in the equation for A_{-j_1} which however can also be handled as explained below.

At a first view there seems to be a full coupling between all equations. However, looking more closely at the coupling terms shows that they have different arguments if $c_j \neq c_n$. We have for example

$$
\begin{aligned}
A_j \left|A_n\right|^2 &= A_j \left(\varepsilon(x - c_j t), \varepsilon^2 t\right) \left|A_n \left(\varepsilon(x - c_n t), \varepsilon^2 t\right)\right|^2 \\
&= A_j(X_j, T) \left|A_n \left(X_j - \varepsilon^{-1}(c_n - c_j)T, T\right)\right|^2 .
\end{aligned}
$$

Hence two spatially localised functions interact only on an $\mathcal{O}(\varepsilon)$-time interval w.r.t. the T-time scale of the NLS equation if $c_j \neq c_n$. Thus the influence of the coupling terms on the dynamics of the NLS equations is only $\mathcal{O}(\varepsilon)$. Hence for spatially localised solutions the NLS equations decouple and the dynamics of the modulations of the carrier waves can be computed for each carrier wave individually by solving

$$
2i\omega_j \partial_T A_j = (1 - c_j^2)\partial_{X_j}^2 A_j - 3A_j \left|A_j\right|^2 .
$$

The argument that spatially localised wave packets with different group velocities do not interact in lowest order has been made rigorous in case of the above NLS approximation first in [21]. The idea has been generalised in [4] for the interaction of various wave packets. It has also been used in case of the interaction of long waves [16, 31, 32]. In case of dissipative systems mean-field coupled Ginzburg–Landau equations take the role of the NLS equation [23].

The interaction of pulse solutions to different carrier waves can be described very precisely such that [21] and [4] can be improved strongly. In [9] and [30] a shift of the underlying carrier wave and a shift of the envelope both of order $\mathcal{O}(\varepsilon)$ are described via the more detailed ansatz

$$
\varepsilon \psi_{\mathrm{multiNLS}} = \sum_{j=1}^{N} \varepsilon A_j \left(\varepsilon(x - c_j t - \varepsilon \psi_j(x, t)), \varepsilon^2 t\right) e^{i(k_j x + \omega_j t + \varepsilon \Omega_j(x,t))} + \mathrm{c.c.},
$$

where the A_j satisfy decoupled equations and where for the pulse shift $\varepsilon \psi_j$ and the phase shift $\varepsilon \Omega_j$ we have explicit formulae. The internal dynamics of the wave packets described via the A_j and the interaction dynamics of the wave packets described via ψ_j and Ω_j can be separated up to very high order. An almost complete description of the interaction of general localised NLS described wave packets can be found in the analytic works [8] and [7].

Remark 5.7.1. In nonlinear optics very often time and space are interchanged. Due to the finite size of the fibers and due to the experimental data which can be measured initial conditions are posed at one end of the fiber, namely at $x = 0$, and one is interested in the solution at the end of the fiber, namely at $x = x_e$,

i. e. x is considered as evolutionary variable, and t as unbounded variable. From a mathematical point of view there is no difference if no dissipation is considered. However, very often phenomenologically dissipation is added to the NLS equation. It has been explained in [29] that it is highly problematic if x and t is interchanged. In this case the NLS equation with dissipation cannot be derived and justified from Maxwell's equation!

5.7.2 Photonic crystals as optical storage

In principle photonic crystals can be used as optical storage. Due to their periodic structure the spectrum in general is not connected (see Ch. 3 for more details) and so called gap solitons can occur. These gap solitons appear at some band edge where the curves of eigenvalues drawn over the Bloch wave numbers have horizontal tangentials, i. e., the group velocity of the wave packet modulated by the gap soliton vanishes. See Figure 5.7. Hence, in principle standing light pulses are possible. In experiments this fact is used to decelerate light already by a factor $1/1000$. These standing light pulses can be found by an approximation via an NLS equation. There exists an approximation result [6] for the NLS approximation in periodic media guaranteeing that these standing light pulses exist at least on time scales of order $\mathcal{O}(1/\varepsilon^2)$. A more detailed analysis [18] using spatial dynamics and invariant manifold theory allows to construct these standing light pulses on time scales of order $\mathcal{O}(1/\varepsilon^n)$ where the size of n depends on the validity of some non-resonance condition. We leave these analytic questions untouched and restrict ourselves to the derivation of the NLS equation. As before a nonlinear wave equation

$$\partial_t^2 u = \chi_1 \partial_x^2 u - \chi_2 u - \chi_3 u^3 \tag{5.29}$$

but now with 2π-periodic coefficient functions χ_1, χ_2 and χ_3, serves as concrete example of an original system. We follow the approach of Section 5.6 by replacing Fourier transform by Bloch wave transform.

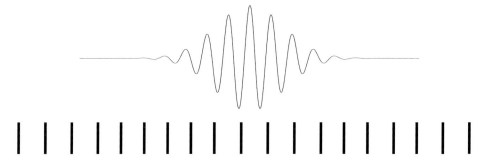

Figure 5.6: A standing light pulse (line) in a photonic crystal (solid bars). The wavelength of the carrier wave and of the photonic crystal are of a comparable order.

Bloch wave transform: Bloch wave transform is defined through

$$\tilde{u}(\ell, x) = (\mathcal{T}u)(\ell, x) = \sum_{j \in \mathbb{Z}} e^{ijx} \hat{u}(\ell + j),$$
$$u(x) = (\mathcal{T}^{-1})\tilde{u}(x) = \int_{-1/2}^{1/2} e^{i\ell x} \tilde{u}(\ell, x) \, d\ell, \tag{5.30}$$

where

$$\tilde{u}(\ell, x) = \tilde{u}(\ell, x + 2\pi) \quad \text{and} \quad \tilde{u}(\ell, x) = \tilde{u}(\ell + 1, x) e^{ix}. \tag{5.31}$$

Bloch wave transform is an isomorphism between

$$H^s(\mathbb{R}, \mathbb{C}) \quad \text{and} \quad L^2((-1/2, 1/2], H^s([0, 2\pi), \mathbb{C})).$$

The multiplication in x-space $u(x)v(x)$ corresponds in Bloch space to the operation

$$(\tilde{u} \star \tilde{v})(\ell, x) = \int_{-\frac{1}{2}}^{\frac{1}{2}} \tilde{u}(\ell - m, x)\tilde{v}(m, x) \, dm, \tag{5.32}$$

where (5.31) has to be used for $|\ell - m| > 1/2$. For 2π periodic $\chi : \mathbb{R} \to \mathbb{R}$ we have $\mathcal{T}(\chi u)(\ell, x) = \chi(x)(\mathcal{T}u)(\ell, x)$. Applying the Bloch wave transform to (5.29) gives

$$\partial_t^2 \tilde{u}(\ell, x) = -\tilde{L}(\ell, \partial_x)\tilde{u}(\ell, x) - \chi_3(x)\tilde{u}^{\star 3}(\ell, x), \tag{5.33}$$

where the operators $\tilde{L}(\ell, \partial_x) : H^2([0, 2\pi)) \to L^2([0, 2\pi))$ are given by

$$\tilde{L}(\ell, \partial_x)\tilde{u}(\ell, \cdot)(x) = -\chi_1(x)(\partial_x + i\ell)^2 \tilde{u}(\ell, x) + \chi_2(x)\tilde{u}(\ell, x).$$

Spectral properties: For fixed ℓ these operators are self adjoint and positive definite in the space $L^2_{\chi_1}([0, 2\pi), \mathbb{C})$ equipped with the inner product

$$\langle \tilde{u}(\ell, \cdot), \tilde{v}(\ell, \cdot) \rangle_{\chi_1} = \int_0^{2\pi} \tilde{u}(\ell, x)\overline{\tilde{v}}(\ell, x)\chi_1(x)^{-1} \, dx.$$

The induced norm $\| \cdot \|_{L^2_{\chi_1}}$ and the usual L^2 norm are equivalent if we assume $\chi_1(x) \geq \gamma > 0$ for a constant γ independent of x by assumption. The self-adjointness of $\tilde{L}(\ell, \partial_x)$ follows from

$$\langle \tilde{L}(\ell, \partial_x)\tilde{u}, \tilde{v} \rangle_{\chi_1} = \int_0^{2\pi} \frac{1}{\chi_1(x)}(-\chi_1(x)(\partial_x + i\ell)^2 \tilde{u}(x) + \chi_2(x)\tilde{u}(x))\overline{\tilde{v}(x)} \, dx$$

$$= \int_0^{2\pi} (-(\partial_x + i\ell)\tilde{u}(x))\overline{(-(\partial_x + i\ell)\tilde{v}(x))} + \frac{\chi_2(x)}{\chi_1(x)}\tilde{u}(x)\overline{\tilde{v}(x)} \, dx$$

$$= \langle \tilde{u}, \tilde{L}(\ell, \partial_x)\tilde{v} \rangle_{\chi_1},$$

and the positive definiteness from

$$\langle \tilde{L}(\ell, \partial_x)\tilde{u}, \tilde{u} \rangle_{\chi_1} = \int_0^{2\pi} |\partial_x \tilde{u}(x)|^2 + \left(\ell^2 + \frac{\chi_2(x)}{\chi_1(x)} \right) |\tilde{u}(x)|^2 \, dx > 0$$

for $\tilde{u} \neq 0$ in $L^2_{\chi_1}$. Thus, for each fixed ℓ there exists a Schauder basis $(f_j(\ell, \cdot))_{j \in \mathbb{N}}$ of $L^2([0, 2\pi))$ of eigenfunctions of $\tilde{L}(\ell, \partial_x)$ with strictly positive eigenvalues $\omega_j^2(\ell)$, i. e. $\tilde{L}(\ell, \partial_x) f_j(\ell, \cdot) = \omega_j^2(\ell) f_j(\ell, \cdot)$. See Figure 5.7.

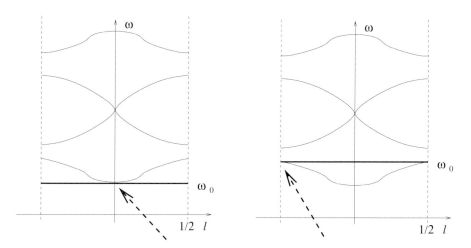

Figure 5.7: The curves of eigenvalues drawn over the Bloch wave numbers ℓ. There are horizontal tangentials at the wave numbers $\ell = 0, \pm 1/2$.

The diagonalised system: Since $\tilde{L}(\ell, \partial_x)$ is self-adjoint in $L^2_{\chi_1}$, the eigenfunctions $(f_j(\ell, \cdot))_{j \in \mathbb{N}}$ can be chosen to form an orthonormal basis of $L^2_{\chi_1}$ for each fixed ℓ. Moreover, the $(f_j(\ell, \cdot))$ can be chosen to be smooth w.r.t. ℓ. We expand the solution for fixed ℓ with respect to this orthogonal basis, i. e.

$$\tilde{u}(\ell, x, t) = \sum_{j \in \mathbb{N}} \tilde{u}_j(\ell, t) f_j(\ell, x),$$

with $\tilde{u}_j(\ell, t) = \tilde{u}_j(\ell + 1, t)$ due to (5.31). The coefficient functions

$$\tilde{u}_j(\ell, t) = \langle f_j(\ell, \cdot), \tilde{u}(\ell, \cdot, t) \rangle_{\chi_1}$$

satisfy

$$\partial_t^2 \tilde{u}_j(\ell, t) = -\omega_j^2 \tilde{u}_j(\ell, t) + \langle f_j(\ell, \cdot), \chi_3(\cdot) \tilde{u}^{\star 3}(\ell, \cdot, t) \rangle_{\chi_1}. \tag{5.34}$$

By introducing

$$\tilde{v}_j(\ell) = \partial_t \tilde{u}_j(\ell)/\omega_j(\ell), \quad \tilde{v}(\ell) = (\tilde{v}_j(\ell))_{j \in \mathbb{N}},$$

where $\omega_{-j}(\ell) = -\omega_j(\ell)$ we rewrite these second order equations as first order systems, for $j \in \mathbb{Z}$. These systems are diagonalised for fixed ℓ. Again by chance the transform is independent of ℓ. We set

$$\tilde{Z}_j = \begin{pmatrix} \tilde{z}_j \\ \tilde{z}_{-j} \end{pmatrix} = \hat{U}^* \begin{pmatrix} \tilde{u}_j \\ \tilde{v}_j \end{pmatrix}, \quad \text{where} \quad \hat{U} = \frac{1}{\sqrt{2}} \begin{pmatrix} 1 & 1 \\ i & -i \end{pmatrix}, \quad \hat{U}^* = \frac{1}{\sqrt{2}} \begin{pmatrix} 1 & -i \\ 1 & i \end{pmatrix}.$$

where as above $\tilde{Z}_j(\ell, t) = \tilde{Z}_j(\ell + 1, t)$, due to $\tilde{u}_j(\ell, t) = \tilde{u}_j(\ell + 1, t)$ and $\omega_j(\ell) = \omega_j(\ell + 1)$. We obtain

$$\partial_t \tilde{Z}_j(\ell, t) = \Lambda_j(\ell) \tilde{Z}_j(\ell, t) - \hat{U}^* \begin{pmatrix} 0 \\ \tilde{s}_j(\ell, t) \end{pmatrix}, \tag{5.35}$$

with $\tilde{\Lambda}_j(\ell) = \mathrm{diag}(i\omega_j(\ell), i\omega_{-j}(\ell))$ and

$$
\begin{aligned}
\tilde{s}_j(\ell, t) &= \int_{-\frac{1}{2}}^{\frac{1}{2}} \int_{-\frac{1}{2}}^{\frac{1}{2}} \sum_{j_1, j_2, j_3 \in \mathbb{N}} \tilde{\beta}^j_{j_1, j_2, j_3}(\ell, \ell-\ell_1, \ell_1-\ell_2, \ell_2)(\tilde{z}_{j_1} + \tilde{z}_{-j_1})(\ell-\ell_1, t) \\
&\quad \times (\tilde{z}_{j_2} + \tilde{z}_{-j_2})(\ell_1 - \ell_2, t)(\tilde{z}_{j_3} + \tilde{z}_{-j_3})(\ell_2, t) \, \mathrm{d}\ell_2 \, \mathrm{d}\ell_1, \\
\tilde{\beta}^j_{j_1, j_2, j_3} &= \frac{1}{\sqrt{8}} \omega_j^{-1}(\ell) \langle f_j(\ell, \cdot), \chi_3(\cdot) f_{j_1}(\ell - \ell_1, \cdot) f_{j_2}(\ell_1 - \ell_2, \cdot) f_{j_3}(\ell_1, \cdot) \rangle_{\chi_1}.
\end{aligned}
$$

System (5.35) is of the form

$$\partial_t \tilde{z}_j(\ell, t) = i\omega_j(\ell) \tilde{z}_j(\ell, t) + \text{ nonlinear terms} \tag{5.36}$$

for j in some index set with nonlinear terms possessing a convolution structure as in the Fourier case of the previous sections. Since (5.36) has the same structure as (5.28) the NLS equation is also the universal amplitude equation describing slow modulations in time and space of a propagating wave packet in spatially periodic dispersive systems, as they occur in the description of nonlinear optics in photonic crystals.

At the band edges we have group velocity zero. The NLS approximation therefore gives standing pulses in lowest order. As already said it has been shown rigorously in [18] that such solutions exist on time scales $\mathcal{O}(1/\varepsilon^n)$ where the size n depends on the validity of some non-resonance condition.

5.7.3   Photonic crystals as wave guides

Photonic crystals can be used to guide light. See Figure 5.8. The periodic structure perpendicular to the direction of the pulse propagation creates band gaps and hinders light of certain temporal wave numbers to emerge perpendicular to the direction of the pulse. Hence, a photonic crystal with a defect, i. e. with an additional structure that breaks the periodicity, can guide light along the defect, thus creating a wave guide.

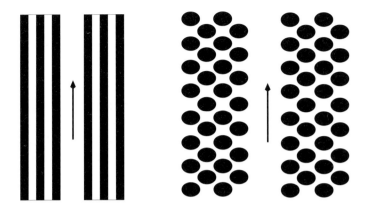

Figure 5.8: Wave guides in one dimensional and two dimensional photonic crystals.

The model problem: As an example we consider a nonlinear wave equation but now in $\mathbb{R}^2$, namely

$$\partial_t^2 u = \Delta u - V u + \gamma u^3, \tag{5.37}$$

with $u = u(x, y, t) \in \mathbb{R}$, $(x, y) \in \mathbb{R}^2$, $t \in \mathbb{R}$, $\gamma \in \{-1, 1\}$, and $V = V(y) = V_{\mathrm{loc}}(y) + V_{\mathrm{per}}(y)$ with $V_{\mathrm{per}}(y) = V_{\mathrm{per}}(y + 2\pi)$ and

$$|V_{\mathrm{loc}}(y)| \leq C e^{-\beta |y|} \text{ for some } C, \beta > 0.$$

See Figure 5.9. It is the purpose of this section to derive a Nonlinear Schrödinger equation describing the pulse propagation along the x-axis, i. e. along a defect of the photonic crystal described by V_{loc} and V_{per}.

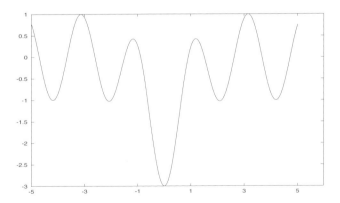

Figure 5.9: A possible function $y \mapsto V(y)$.

The linearised problem: As a first step we consider the linearised problem

$$\partial_t^2 u = \Delta u - V u. \tag{5.38}$$

We look for solutions $u(x, y, t) = e^{i\omega t} e^{ikx} v(y)$ and find that v has to satisfy

$$v'' - k^2 v - V v = -\omega^2 v \qquad \text{or} \qquad -L v = -v'' + V v = (\omega^2 - k^2) v.$$

In case $V_{\text{loc}} = 0$ the operator L possesses spectral gaps, i. e.,

$$\sigma(-L) = \bigcup_{j \in \mathbb{N}} [\alpha_j, \beta_j]$$

with $\ldots \le \alpha_j < \beta_j \le \alpha_{j+1} < \beta_{j+1} \le \ldots$. In case $V_{\text{loc}} \ne 0$ discrete eigenvalues can emerge from the spectral bands into the spectral gaps. See Figure 5.10. We assume that there is at least one such eigenvalue λ with $\beta_j < \lambda < \alpha_{j+1}$. This λ is the eigenvalue associated to the wave guide.

Figure 5.10: A possible spectrum $\sigma(-L)$.

The eigenfunction associated to λ is called φ, i. e., $L\varphi = -\lambda\varphi$. The adjoint eigenfunction is called φ^* and satisfies $\langle \varphi^*, \varphi \rangle = 1$. The spectral values of the linearisation $\Delta u - V u$ are given by $\sigma(L) - k^2$, i. e., the spectral values can be drawn as a function over the Fourier wave numbers $k \in \mathbb{R}$. See Figure 5.11. The curve $k \mapsto -\lambda - k^2$ will be the one for which we derive the NLS equation. Therefore, this NLS equation will describe wave packets which move along the defect, i. e., along the x-axis.

Derivation of the NLS equation: We split u into modes associated to φ and into the rest, i. e., we set $u = c\varphi + w$ with $\langle \varphi^*, w \rangle = 0$ and find

$$\partial_t^2 c = (-\lambda + \partial_x^2)c + \gamma \langle \varphi^*, (c\varphi + w)^3 \rangle, \tag{5.39}$$
$$\partial_t^2 w = \Delta w + V w + \gamma \langle 1 - \varphi^*, (c\varphi + w)^3 \rangle. \tag{5.40}$$

In order to derive the NLS equation we make the ansatz

$$c = \varepsilon A_1 \left(\varepsilon(x - c_g t), \varepsilon^2 t \right) e^{i(k_0 x + \omega_0 t)} + \text{c.c.}, \tag{5.41}$$

with $A_1(X, T) \in \mathbb{C}$. At $\mathcal{O}(\varepsilon)$ and $\mathcal{O}(\varepsilon^2)$ in the c-equation we find the linear dispersion relation and the group velocity for the wave packet

$$\omega_0^2 = \lambda_0 + k_0^2, \qquad \text{and} \qquad c_g = \frac{d\omega}{dk}\Big|_{k=k_0} = \frac{k_0}{\omega_0}.$$

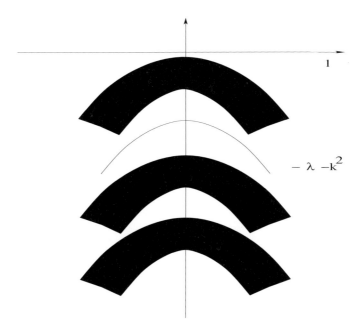

Figure 5.11: A possible spectrum of $\sigma(L - k^2 I)$ drawn over the Fourier wave numbers k.

At $\mathcal{O}(\varepsilon^3)$ in the c-equation we find the NLS equation

$$2\mathrm{i}\omega_0 \partial_T A_1 = \partial_X^2 A_1 + 3\gamma \langle \varphi^*, \varphi |\varphi|^2 \rangle A_1 |A_1|^2 . \tag{5.42}$$

We expect that under relatively weak assumptions the following approximation result holds.

Claim 5.7.2. *Let $A \in C([0, T_0], H^{s_A}(\mathbb{R}, \mathbb{C}))$ be a solution of the NLS equation (5.42) with s_A being sufficiently large. Then there exist $\varepsilon_0 > 0$ and $C > 0$ such that for all $\varepsilon \in (0, \varepsilon_0)$ the difference between true solutions of (5.37) and the approximation defined in (5.41) is less than $C\varepsilon^2$ in the sup-norm uniformly for all $t \in [0, T_0/\varepsilon^2]$.*

Outline of a proof: The major difficulty lies in the construction of an approximation $\varepsilon\psi$ which makes the residual

$$\mathrm{Res}(u) = -\partial_t^2 u + \Delta u - V u + \gamma u^3$$

small. If this smallness is established, the error estimates easily follows with a simple application of Gronwall's inequality. In order to show this, suppose that we have an approximation with $\varepsilon^{-2}\mathrm{Res}(\varepsilon\psi) = \mathcal{O}(\varepsilon^2)$. The error $\varepsilon^2 R = u - \varepsilon\psi$ made by this approximation satisfies

$$\partial_t^2 R = \Delta R - V R + \gamma \left(3\varepsilon^2 \psi^2 R + 3\varepsilon^2 \psi R^2 + \varepsilon^4 R^3 + \varepsilon^{-2}\mathrm{Res}(\varepsilon\psi) \right) .$$

We consider the energy

$$E = \int \left((\partial_t R)^2 + (\nabla R)^2 + V R^2 \right) dx$$

and find

$$
\begin{aligned}
\frac{1}{2}\partial_t E &= \int \left((\partial_t R)\left(\Delta R - VR + \gamma\left(3\varepsilon^2\psi^2 R + 3\varepsilon^3\psi R^2 + \varepsilon^4 R^3 + \varepsilon^{-2}\mathrm{Res}(\varepsilon\psi)\right)\right) \right. \\
&\qquad \left. - (\partial_t R)\Delta R + \partial_t R V R \right) dx \\
&= \int \left((\partial_t R)\gamma\left(3\varepsilon^2\psi^2 R + 3\varepsilon^3\psi R^2 + \varepsilon^4 R^3 + \varepsilon^{-2}\mathrm{Res}(\varepsilon\psi)\right) \right) dx
\end{aligned}
$$

and so

$$|\partial_t E| \le C_1(\psi)\varepsilon^2 E + C_2(\psi)\varepsilon^3 E^{3/2} + \varepsilon^4 E^2 + C_{\mathrm{Res}}\varepsilon^2.$$

Since $E(0) = 0$ Gronwall's inequality implies for $\varepsilon > 0$ sufficiently small that $\sup_{t \in [0, T_0/\varepsilon^2]} E(t) = \mathcal{O}(1)$. Since $\sqrt{E(t)}$ is equivalent to the H^1 norm if $\inf_{x \in \mathbb{R}} V(x) > 0$ we have shown $\sup_{t \in [0, T_0/\varepsilon^2]} \|R(t)\|_{H^1} = \mathcal{O}(1)$.

Smallness of the residual: In order to make the residual small additional lower order terms are added to the approximation. There are two fundamentally different situations, namely when integer multiples of ω_0 fall into spectral gaps and when integer multiples of ω_0 do not fall into spectral gaps. The second case is much more involved and is left to future research. In the first case we make the ansatz $\varepsilon\psi = (\varepsilon\psi_c, \varepsilon\psi_w)$ with

$$
\begin{aligned}
\varepsilon\psi_c &= \varepsilon A_1\left(\varepsilon(x - c_g t), \varepsilon^2 t\right) e^{i(k_0 x + \omega_0 t)} + \text{c.c.}, \\
&\quad + \varepsilon^3 A_3\left(\varepsilon(x - c_g t), \varepsilon^2 t\right) e^{3i(k_0 x + \omega_0 t)} + \text{c.c.}, \\
\varepsilon\psi_w &= \varepsilon^3 W_1\left(\varepsilon(x - c_g t), \varepsilon^2 t\right) e^{i(k_0 x + \omega_0 t)} + \text{c.c.} \\
&\quad + \varepsilon^3 W_3\left(\varepsilon(x - c_g t), \varepsilon^2 t\right) e^{3i(k_0 x + \omega_0 t)} + \text{c.c.},
\end{aligned}
$$

and define $A_{-j} = \overline{A_j}$ and $W_{-j} = \overline{W_j}$. Inserting this approximation into (5.39)-(5.40) gives the following set of equations

$$
\begin{aligned}
2i\omega_0 \partial_T A_1 &= \partial_X^2 A_1 + 3\gamma\langle \varphi^* | \varphi\, |\varphi|^2 \rangle A_1 |A_1|^2, \\
-(3\omega_0)^2 A_3 &= \left(-\lambda_0 - (3k_0)^2\right) A_3 + \gamma\langle \varphi^*, \varphi^3 \rangle A_1^3, \\
-(\omega_0)^2 W_1 &= \left(\partial_y^2 - k_0^2 - V\right) W_1 + 3\gamma\langle 1 - \varphi^*, \varphi\, |\varphi|^2 \rangle A_1 |A_1|^2, \\
-(3\omega_0)^2 W_3 &= \left(\partial_y^2 - (3k_0)^2 - V\right) W_3 + \gamma\langle 1 - \varphi^*, \varphi^3 \rangle A_1^3.
\end{aligned}
\tag{5.43}
$$

We made this ansatz under the assumption that integer multiples of ω_0 falls into spectral gaps, i. e., we assume

(S1) $-(3\omega_0)^2 + \lambda_0 + (3k_0)^2) \ne 0,$

(S2) $-\omega_0^2 \notin \sigma(\partial_y^2 - k_0^2 - V),$

(S3) $-(3\omega_0)^2 \notin \sigma(\partial_y^2 - (3k_0)^2 - V).$

The second equation can be solved with respect to A_3 if **(S1)** is satisfied. The third equation can be resolved with respect to W_3 if **(S2)** is satisfied. The fourth equation can be resolved with respect to W_1 if **(S3)** is satisfied. Hence, under the assumptions **(S1)**–**(S3)** the residual $\mathrm{Res}(\varepsilon\psi_c, \varepsilon\psi_s)$ is formally of order $\mathcal{O}(\varepsilon^4)$ if the above ansatz is plugged in. Due to the loss of ε^{-1} by scaling in $\mathbb{R}^2$ the residual is only $\mathcal{O}(\varepsilon^3)$ in L^2 which is not sufficient according to the above outline of a proof.

In order to obtain $\mathcal{O}(\varepsilon^4)$ in L^2 we have to extend the above approximation further, namely $\varepsilon\psi_w$ by

$$+ \varepsilon^4 W_{11}\left(\varepsilon(x - c_g t), \varepsilon^2 t\right) e^{\mathrm{i}(k_0 x + \omega_0 t)} + \text{c.c.}$$
$$+ \varepsilon^4 W_{31}\left(\varepsilon(x - c_g t), \varepsilon^2 t\right) e^{3\mathrm{i}(k_0 x + \omega_0 t)} + \text{c.c.},$$

and $\varepsilon\psi_c$ by

$$\varepsilon^2 A_{11}\left(\varepsilon(x - c_g t), \varepsilon^2 t\right) e^{\mathrm{i}(k_0 x + \omega_0 t)} + \text{c.c.},$$
$$+ \varepsilon^4 A_{31}\left(\varepsilon(x - c_g t), \varepsilon^2 t\right) e^{3\mathrm{i}(k_0 x + \omega_0 t)} + \text{c.c.},$$

The variable A_{11} solves an inhomogeneous linear Schrödinger equation and the variables A_{31}, W_{11}, and W_{31} satisfy algebraic equations which can be solved under the same conditions **(S1)**–**(S3)**. Hence we have established the validity of the above claim under the validity of the non-resonance conditions **(S1)**–**(S3)**.

Theorem 5.7.3. *Under the assumptions* **(S1)**–**(S3)** *the following holds. Let $A \in C([0, T_0], H^{s_A}(\mathbb{R}, \mathbb{C}))$ be a solution of the NLS equation* (5.42) *with s_A being sufficiently large. Then there exist $\varepsilon_0 > 0$ and $C > 0$ such that for all $\varepsilon \in (0, \varepsilon_0)$ the difference between true solutions of* (5.37) *and the approximation is less than $C\varepsilon^2$ in the sup-norm uniformly for all $t \in [0, T_0/\varepsilon^2]$.*

We strongly expect that the method presented in this section applies to Maxwell's equations and two dimensional photonic crystals, i. e., when Fourier transform in the x-direction has to be replaced by Bloch wave transform. This has to be subject of future research.

Bibliography

[1] G. P. Agrawal. *Nonlinear fiber optics.* Academic Press, 3rd edition, 2001.

[2] H.-W. Alt. *Linear functional analysis. An application oriented introduction. (Lineare Funktionalanalysis. Eine anwendungsorientierte Einführung.) 5th revised ed.* Berlin: Springer, 2006.

[3] M. J. Ablowitz and H. Segur. *Solitons and the inverse scattering transform,* volume 4 of *SIAM Studies in Applied Mathematics.* Society for Industrial and Applied Mathematics (SIAM), Philadelphia, Pa., 1981.

[4] A. Babin and A. Figotin. Linear superposition in nonlinear wave dynamics. *Rev. Math. Phys.*, 18(9):971–1053, 2006.

[5] J. Bourgain. *Global solutions of nonlinear Schrödinger equations*, volume 46 of *American Mathematical Society Colloquium Publications*. American Mathematical Society, Providence, RI, 1999.

[6] K. Busch, G. Schneider, L. Tkeshelashvili, and H. Uecker. Justification of the nonlinear Schrödinger equation in spatially periodic media. *Z. Angew. Math. Phys.*, 57(6):905–939, 2006.

[7] M. Chirilus-Bruckner. *Nonlinear Interaction of Pulses*. PhD thesis, TU Karlsruhe, 2009.

[8] M. Chirilus-Bruckner, C. Chong, G. Schneider, and H. Uecker. Separation of internal and interaction dynamics for NLS-described wave packets with different carrier waves. *J. Math. Anal. Appl.*, 347(1):304–314, 2008.

[9] M. Chirilus-Bruckner, G. Schneider, and H. Uecker. On the interaction of NLS-described modulating pulses with different carrier waves. *Math. Methods Appl. Sci.*, 30(15):1965–1978, 2007.

[10] P. G. Drazin and R. S. Johnson. *Solitons: an introduction*. Cambridge Texts in Applied Mathematics. Cambridge University Press, Cambridge, 1989.

[11] W.-P. Düll and G. Schneider. Justification of the nonlinear Schrödinger equation for a resonant Boussinesq model. *Indiana Univ. Math. J.*, 55(6):1813–1834, 2006.

[12] J. D. Gibbon. Why the NLS equation is simultaneously a success, a mediocrity and a failure in the theory of nonlinear waves. In *Soliton theory: a survey of results*, Nonlinear Sci. Theory Appl., pages 133–151. Manchester Univ. Press, Manchester, 1990.

[13] T. Gallay and G. Schneider. KP description of unidirectional long waves. The model case. *Proc. Roy. Soc. Edinburgh Sect. A*, 131(4):885–898, 2001.

[14] M. Hisakado. Breather trapping mechanism in piecewise homogeneous dna. *Physics Letters A*, 227:87–93, 1997.

[15] L. A. Kalyakin. Asymptotic decay of a one-dimensional wave packet in a nonlinear dispersive medium. *Mat. Sb. (N.S.)*, 132(174)(4):470–495, 592, 1987.

[16] L. A. Kalyakin. Long-wave asymptotics. Integrable equations as the asymptotic limit of nonlinear systems. *Uspekhi Mat. Nauk*, 44(1(265)):5–34, 247, 1989.

[17] P. Kirrmann, G. Schneider, and A. Mielke. The validity of modulation equations for extended systems with cubic nonlinearities. *Proc. Roy. Soc. Edinburgh Sect. A*, 122(1-2):85–91, 1992.

[18] V. Lescarret, C. Blank, M. Chirilus-Bruckner, C. Chong, and G. Schneider. Standing generalized modulating pulse solutions for a nonlinear wave equation in periodic media. *Nonlinearity*, 22:1869–1898, 2009.

[19] H. Leblond. Direct derivation of a macroscopic NLS equation from the quantum theory. *J. Phys. A: Math. Gen.*, 34:3109–3123, 2001.

[20] A. Pazy. *Semigroups of linear operators and applications to partial differential equations*, volume 44 of *Applied Mathematical Sciences*. Springer-Verlag, New York, 1983.

[21] R. D. Pierce and C. E. Wayne. On the validity of mean-field amplitude equations for counterpropagating wavetrains. *Nonlinearity*, 8(5):769–779, 1995.

[22] G. Schneider. Validity and limitation of the Newell–Whitehead equation. *Math. Nachr.*, 176:249–263, 1995.

[23] G. Schneider. Justification of mean-field coupled modulation equations. *Proc. Roy. Soc. Edinburgh Sect. A*, 127(3):639–650, 1997.

[24] G. Schneider. Approximation of the Korteweg–de Vries equation by the nonlinear Schrödinger equation. *J. Differential Equations*, 147(2):333–354, 1998.

[25] G. Schneider. Justification of modulation equations for hyperbolic systems via normal forms. *NoDEA Nonlinear Differential Equations Appl.*, 5(1):69–82, 1998.

[26] G. Schneider. Justification and failure of the nonlinear Schrödinger equation in case of non-trivial quadratic resonances. *J. Differential Equations*, 216(2):354–386, 2005.

[27] C. Sulem and P.-L. Sulem. *The nonlinear Schrödinger equation*, volume 139 of *Applied Mathematical Sciences*. Springer-Verlag, New York, 1999. Self-focusing and wave collapse.

[28] W. A. Strauss. *Nonlinear wave equations*, volume 73 of *CBMS Regional Conference Series in Mathematics*. Published for the Conference Board of the Mathematical Sciences, Washington, DC, 1989.

[29] G. Schneider and H. Uecker. Existence and stability of modulating pulse solutions in Maxwell's equations describing nonlinear optics. *Z. Angew. Math. Phys.*, 54(4):677–712, 2003.

[30] G. Schneider, H. Uecker, and M. Wand. Interaction of modulated pulses in nonlinear oscillator chains. *Journal of Difference Equations and Applications*, 2009. accepted.

[31] G. Schneider and C. E. Wayne. The long-wave limit for the water wave problem. I. The case of zero surface tension. *Comm. Pure Appl. Math.*, 53(12):1475–1535, 2000.

[32] G. Schneider and C. Eugene Wayne. The rigorous approximation of long-wavelength capillary-gravity waves. *Arch. Ration. Mech. Anal.*, 162(3):247–285, 2002.

[33] T. Tao. *Nonlinear dispersive equations*, volume 106 of *CBMS Regional Conference Series in Mathematics*. Published for the Conference Board of the Mathematical Sciences, Washington, DC, 2006. Local and global analysis.

[34] V. E. Zakharov. Stability of periodic waves of finite amplitude on the surface of a deep fluid. *Sov. Phys. J. Appl. Mech. Tech. Phys,*, 4:190–194, 1968.

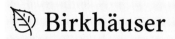
Oberwolfach Seminars

The workshops organized by the *Mathematisches Forschungsinstitut Oberwolfach* are intended to introduce students and young mathematicians to current fields of research. By means of these well-organized seminars, also scientists from other fields will be introduced to new mathematical ideas. The publication of these workshops in the series *Oberwolfach Seminars* (formerly *DMV Seminar*) makes the material available to an even larger audience.

■ **OWS 41: Hacon, C.D. / Kovács, S.**, Classification of Higher Dimensional Algebraic Varieties (2010).
ISBN 978-3-0346-0289-1

This book focuses on recent advances in the classification of complex projective varieties. It is divided into two parts. The first part gives a detailed account of recent results in the minimal model program. In particular, it contains a complete proof of the theorems on the existence of flips, on the existence of minimal models for varieties of log general type and of the finite generation of the canonical ring. The second part is an introduction to the theory of moduli spaces. It includes topics such as representing and moduli functors, Hilbert schemes, the boundedness, local closedness and separatedness of moduli spaces and the boundedness for varieties of general type.
The book is aimed at advanced graduate students and researchers in algebraic geometry.

■ **OWS 40: Baum, H. / Juhl, A.**, Conformal Differential Geometry. Q-Curvature and Conformal Holonomy (2010).
ISBN 978-3-7643-9908-5
Conformal invariants (conformally invariant tensors, conformally covariant differential operators, conformal holonomy groups etc.) are of central significance in differential geometry and physics. Well-known examples of conformally covariant operators are the Yamabe, the Paneitz, the Dirac and the twistor operator. These operators are intimately connected with the notion of Branson's Q-curvature. The aim of these lectures is to present the basic ideas and some of the recent developments around Q-curvature and conformal holonomy. The part on Q-curvature starts with a discussion of its origins and its

relevance in geometry and spectral theory. The subsequent lectures describe the fundamental relation between Q-curvature and scattering theory on asymptotically hyperbolic manifolds. Building on this, they introduce the recent concept of Q-curvature polynomials and use these to reveal the recursive structure of Q-curvatures. The part on conformal holonomy starts with an introduction to Cartan connections and their holonomy groups. Then we define holonomy groups of conformal manifolds, discuss their relation to Einstein metrics and recent classification results in Riemannian and Lorentzian signature. In particular, we explain the connection between conformal holonomy and conformal Killing forms and spinors, and describe Fefferman metrics in CR geometry as Lorentzian manifolds with conformal holonomy $SU(1,m)$.

■ **OWS 39: Drton, M. / Sturmfels, B. / Sullivant, S.**, Lectures on Algebraic Statistics (2008).
ISBN 978-3-7643-8904-8

■ **OWS 38: Bobenko, A.I. / Schröder, P. / Sullivan, J.M. / Ziegler, G.M. (Eds.)**, Discrete Differential Geometry (2008).
ISBN 978-3-7643-8620-7

■ **OWS 37: Galdi, G.P. / Rannacher, R. / Robertson, A.M. / Turek, S.**, Hemodynamical Flows (2008).
ISBN 978-3-7643-7805-9

■ **OWS 36: Cuntz, J. / Meyer, R. / Rosenberg, J.M.**, Topological and Bivariant K-theory (2007).
ISBN 978-3-7643-8398-5

■ **OWS 35: Itenberg, I. / Mikhalkin, G. / Shustin, E.**, Tropical Algebraic Geometry (2007).
ISBN 978-3-7643-8309-1

Printing: Ten Brink, Meppel, The Netherlands
Binding: Stürtz, Würzburg, Germany